Local Flora and Fauna Conservation – a Role for Zoos, Aquariums and Botanical Gardens

Local Flora and Fauna Conservation – a Role for Zoos, Aquariums and Botanical Gardens

Editors

Ursula Bechert
Debra C. Colodner

MDPI • Basel • Beijing • Wuhan • Barcelona • Belgrade • Manchester • Tokyo • Cluj • Tianjin

Editors
Ursula Bechert
School of Arts and Sciences
University of Pennsylvania
Philadelphia
United States

Debra C. Colodner
Conservation Education and
Science Department
Arizona-Sonora Desert Museum
Tucson
United States

Editorial Office
MDPI
St. Alban-Anlage 66
4052 Basel, Switzerland

This is a reprint of articles from the Special Issue published online in the open access journal *Journal of Zoological and Botanical Gardens* (ISSN 2673-5636) (available at: www.mdpi.com/journal/jzbg/special_issues/local_conservation).

For citation purposes, cite each article independently as indicated on the article page online and as indicated below:

LastName, A.A.; LastName, B.B.; LastName, C.C. Article Title. *Journal Name* **Year**, *Volume Number*, Page Range.

ISBN 978-3-0365-7865-1 (Hbk)
ISBN 978-3-0365-7864-4 (PDF)

Contents

About the Editors

Ursula Bechert

Ursula S. Bechert earned her B.S. degree from Utah State University in 1985 and her DVM from Washington State University in 1991. After working in private practice and at a free-ranging wildlife park in Oregon for several years, she earned a Ph.D. in animal sciences from Oregon State University. She has been conducting research for over 30 years, with interests in the reproductive physiology of endangered species, the development of novel population management tools, and nutritional and pharmacokinetic studies with wildlife. She is currently serving as the research advisor for the Bureau of Land Management's Wild Horse and Burro Advisory Board. At the University of Pennsylvania, she helps shape the strategic goals of graduate-level programming for the College of Liberal and Professional Studies in the School of Arts and Sciences.

Debra C. Colodner

Debra Colodner is Director of Conservation Education and Science at the Desert Museum, where she oversees interpretive and educational programs, conservation science and outreach, as well as natural history collections. Debra is an oceanographer by training (Ph.D., MIT) who was lured to Tucson, AZ, by her passion for conservation and science education, though the ocean there is long gone. In Tucson, she dove into formal and informal education, directing education programs at Biosphere 2 (under Columbia University), the Flandrau Science Center at the University of Arizona, and most recently, at the Desert Museum. The Museum's current conservation programs promote biodiversity by developing and sharing a better understanding of its value for people. Desert Museum science and education programs focus on endangered and threatened species, invasive species, pollinators, and conserving the wild relatives of our crops.

Preface to "Local Flora and Fauna Conservation – a Role for Zoos, Aquariums and Botanical Gardens"

Dear colleagues,

Over the past century, zoos, aquariums, and botanical gardens have evolved from serving as places of entertainment to becoming centers for species conservation, primarily through education and research. Numerous factors can contribute to the decline of a species, including habitat loss, resource limitations tied to climate change, competition with introduced or invasive species, and disease. Solutions are often complex and require collaborative efforts across different types of organizations that have different expertise and resources.

Many zoos partner with overseas organizations to help fund research and in-situ conservation based on the exotic species that they exhibit. However, the greatest conservation success stories are those that were conducted close to home and in collaboration with local government agencies and other organizations (e.g., black-footed ferret, California condor). Multiple stakeholders are usually involved in habitat preservation initiatives to balance economic and environmental goals. New policies may be needed to protect key species and landscapes. Educational initiatives can motivate individuals to get involved. All of these activities are more effective when focused on indigenous species at a local level, yet there is very little in the literature about how organizations have successfully exhibited native species and worked collaboratively across multiple organizations on conservation efforts.

This Special Issue reprint highlights how zoos, aquariums, and botanical gardens have partnered with organizations and agencies to address local and regional conservation issues through research and education projects, rehabilitation programs, policymaking efforts, and other types of activities. Understanding how such partnerships are developed and maintained, the challenges encountered, and success stories (as well as unsuccessful ones) were of interest. Hopefully, the stories shared and lessons learned will guide other organizations in their local and regional conservation efforts.

Ursula Bechert and Debra C. Colodner
Editors

Journal of
Zoological and Botanical Gardens

MDPI

Editorial

Regional Conservation, Research, and Education: Ways Forward

Ursula S. Bechert

School of Arts & Sciences, University of Pennsylvania, 3440 Market Street, Suite 450, Philadelphia, PA 19104, USA; bechertu@sas.upenn.edu

Citation: Bechert, U.S. Regional Conservation, Research, and Education: Ways Forward. *J. Zool. Bot. Gard.* **2023**, 4, 288–291. https://doi.org/10.3390/jzbg4010024

Received: 7 March 2023
Accepted: 8 March 2023
Published: 17 March 2023

There are currently over 8 billion people on Earth, a figure which grows by approximately 67 million annually; https://www.worldometers.info/ (accessed on 3 March 2023). As an IUCN report stated over 30 years ago, "The combined distributive impacts of an affluent resource-consuming minority and a poor majority struggling to stay alive are inexorably and rapidly destroying the buffer that has always existed, at least on a global scale, between human resource consumption and the planet's productive capacity" [1]. More and more people are relying on less and less land for sustenance; human social and economic developments are inextricably linked to biological diversity and the health of ecosystems. So, conservation is complicated.

Environmental conservation groups comprise one of the fastest growing nonprofit sectors [2], and among these are some organizations that are in fact antienvironmental in their actions [3]. Groups vie for donor support, and even when significant funding is available, large international conservation organizations often struggle to spend funds on prioritized projects [4]. Failure to understand the human cultural, political, and socioeconomic factors in a particular geographic area can contribute to the failure of conservation initiatives [5]. Transfrontier conservation areas (TFCAs) in Southern Africa conserve biodiversity across country borders by engaging people across all levels of governance (e.g., political leaders, local communities, nongovernmental organizations (NGOs), and private sector) in the process of decision making [6]. However, this process is ongoing as human populations continue to grow, wildlife habitats diminish in size, and human–wildlife conflicts increase in frequency [7]. Conservation reserves increase connectivity across protected areas to allow for the movement of animals for breeding purposes and in response to climate change, and thereby create more resilient ecosystems [8]. Connecting landscapes to preserve biodiversity is the mission of several organizations in the U.S. (e.g., the Wildlands Network) and is a common strategy for some NGOs such as The Nature Conservancy [9]. The federal government's America the Beautiful Initiative supports locally led efforts with the goal of conserving at least 30% of the nation's lands and waters by 2030 [10]. Such large-scale projects require collaboration across a wide range of organizations [11].

This Special Issue explores how zoos, aquaria, and botanical gardens work with different kinds of organizations on local or regional conservation projects. Zoos and aquaria represent some of the biggest nonprofits involved in ex situ conservation [12]; however, local and regional in situ conservation projects are now being pursued by more zoos, aquariums, and botanical gardens than ever before. Unfortunately, many conservation organizations operate independently from one another or with other institutions only on a project-by-project basis [13]. The inability to link different types of organizations across institutional and disciplinary boundaries frequently results in fragmented (e.g., species-specific) or unprioritized research, inefficient resource utilization, duplication, and lack of accountability regarding the long-term continuity of programs or the actual implementation of research results. Existing linkage organizations are primarily sector-specific (e.g., the American Association of Zoos and Aquariums). Most individual research and education institutions have their own libraries and database systems, but many cannot talk to one another, or their datasets are focused within and serve specific disciplines.

Regionally organized cooperatives can serve as instruments for capacity building and disseminate scientific research results beyond traditional publications through integration into educational programs and conservation initiatives. A central organizing body can reduce the transactional costs for collaborating organizations, provide a more competitive funding platform, and improve conservation outcomes [13]. Thoughtfully considering how different partnerships should be structured and managed can contribute to long-lasting relationships and successful projects [14]. Very little research has been carried out on the management of collaborative conservation projects and their outcomes [15]. The Free Roaming Equids and Ecosystem Sustainability Network (FREES) is an example of an organizing body, pulling together disparate organizations in Utah to cooperatively manage free-roaming equid populations as a demonstration project. Similarly, zoos, botanical gardens, and aquaria often partner with a broad range of organizations [14,16,17] to pursue common conservation goals.

Most nonprofit organizations focus conservation efforts on in situ, terrestrial, state-specific projects—primarily with birds, followed by fish and then mammals [12]. Megacharismatic mammal as well as bird and fish conservation projects typically receive more funding compared to reptile, amphibian, and invertebrate projects because these species are favored by the general public and government policy agendas [18]. To counter this, zoos, aquaria, and botanical gardens are helping the public better understand the important roles that lesser-known species play in our ecosystems (e.g., Sonoyta pupfish (*Cyprinodon macularis eremus*) and Quitobaquito spring snail (*Tryonia quitobaquitae*) [17]). Zoo-based projects often focus on the conservation of a single species (e.g., amphibians [19], small mammals [16], and reptiles [20]). They frequently serve as ex situ sites for breeding endangered species, which then get released into the wild through partnerships with other organizations. However, zoos, aquaria, and botanical gardens can also get involved in advocacy, research, restoration, and rehabilitation. The sea otter (*Enhydra lutris*) rehabilitation program, initiated by the Monterey Bay Aquarium in 1984, has grown to involve multiple activities including the coordination of a stranding response, husbandry, research, veterinary care, release, population monitoring, and advocacy to implement policy change [21]. By focusing on the conservation of local or endemic species, zoos, aquaria, and botanical gardens can leverage resources from regional agencies and organizations, engage their staff and local communities in the process, and educate the public about the vital role native species play in their own backyard. Over its 70-year history, the Arizona-Sonora Desert Museum has served as a partner in advocacy, captive breeding and field research, and long-term phenology studies and as the lead for the regional management of invasive grasses; it has also helped to develop a community conservation plan [17].

Evaluation tools can be used to better understand and align an organization's goals with their conservation efforts [22]. How conservation organizations describe their mission and are structured plays a critical role in their ability to conserve biodiversity [23]. The Minnesota Zoo is designated as a state agency, so their conservation projects naturally focus on native species. They have found it relatively easy to partner with other state agencies on conservation initiatives, but project-specific grant funds have a tendency to silo personnel [24]. The North Carolina Zoo is the world's largest natural habitat zoo; they protect and manage over 2000 acres of land in collaboration with other agencies to preserve native species and provide hiking trails for the local community [25]. Collectively, this Special Issue highlights lessons learned and how creative partnerships can result in innovative conservation initiatives that would not otherwise be possible for individual organizations.

Acknowledgments: I thank all the authors who contributed work to this Special Issue, as well as the reviewers and editors who strengthened this publication by contributing their time and expertise.

Conflicts of Interest: The author declares no conflict of interest.

References

1. McNeely, J.A.; Miller, K.R.; Reid, W.V.; Mittermeier, R.A.; Werner, T.B. *Conserving the World's Biological Diversity*; WRI, 1709 New York Ave., N.W., Washington, D.C. 20006; CI, 2011 Crystal Dr #600, Arlington, VA 22202; WWF-US, 1250 Twenty-Fourth Street, N.W. P.O. Box 97180 Washington, DC 20090-7180; and the World Bank, 2121 Pennsylvania Avenue, NW, Washington, DC, USA; IUCN: Gland, Switzerland, 1990; pp. 1–185.
2. Straughan, B.; Pollack, T. *The Broader Movement: Nonprofit Environmental and Conservation Organizations 1989–2005*; National Center for Charitable Statistics: Washington, DC, USA, 2008.
3. Stein, J.; Beckel, M. A Guide to Environmental Non-Profits. Mother Jones 2006: Mar/Apr Issue. Available online: https://www.motherjones.com/environment/2006/03/guide-environmental-non-profits/ (accessed on 23 February 2023).
4. Halpern, B.S.; Pyke, C.R.; Fox, H.E.; Haney, C.; Schlaepfer, M.A.; Zaradic, P. Gaps and mismatches between global conservation priorities and spending. *Conserv. Biol.* **2006**, *20*, 56–64. [CrossRef] [PubMed]
5. Bennett, N.J.; Roth, R.; Klain, S.C.; Chan, K.; Christie, P.; Clark, D.A.; Cullman, G.; Curran, D.; Durbin, T.J.; Epstein, G.; et al. Conservation social science: Understanding and integrating human dimensions to improve conservation. *Biol. Conserv.* **2017**, *205*, 93–108. [CrossRef]
6. Hanks, J. Transfrontier conservation areas (TFCAs) in southern Africa. *J. Sust. Forest.* **2003**, *17*, 127–148. [CrossRef]
7. Stoldt, M.; Gottert, T.; Mann, C.; Zeller, U. Transfrontier conservation areas and human-wildlife conflict: The case of the Namibian component of the Kavango-Zambezi (KAZA) TFCA. *Sci. Rep.* **2020**, *10*, 7964. [CrossRef] [PubMed]
8. Belote, R.R.; Dietz, M.S.; Jenkins, C.N.; McKinley, P.S.; Irwin, G.H.; Fullman, T.J.; Leppi, J.C.; Aplet, G.H. Wild, connected, and diverse: Building a more resilient system of protected areas. *Ecol. Applic.* **2017**, *27*, 1050–1056. [CrossRef]
9. Davies, Z.G.; Kareiva, P.; Armsworth, P.R. Temporal patterns in the size of conservation land transactions. *Conserv. Lett.* **2010**, *3*, 29–37. [CrossRef]
10. Fact Sheet: Biden-Harris Administration Celebrates Expansion of Locally-Led Conservation Efforts in First Year of "America the Beautiful" Initiative. 20 December 2021. Available online: https://www.whitehouse.gov/briefing-room/statements-releases/2021/12/20/fact-sheet-biden-harris-administration-celebrates-expansion-of-locally-led-conservation-efforts-in-first-year-of-america-the-beautiful-initiative/ (accessed on 22 December 2022).
11. Bode, M.; Probert, W.; Turner, W.R.; Wilson, K.A.; Venter, O. Conservation planning with multiple organizations and objectives. *Conserv. Biol.* **2010**, *25*, 295–304. [CrossRef] [PubMed]
12. Armsworth, P.R.; Fishburn, I.S.; Davies, Z.G.; Gilbert, J.; Leaver, N.; Gaston, K.J. The size, concentration, and growth of biodiversity-conservation nonprofits. *BioScience* **2012**, *62*, 271–281.
13. Mace, G.M.; Balmford, A.; Boitani, L.; Cowlishaw, G.; Dobson, A.P.; Faith, D.P.; Gaston, K.J.; Humphries, C.J.; Vane-Wright, R.I.; Williams, P.H.; et al. It's time to work together and stop duplicating conservation efforts. *Nature* **2000**, *405*, 393. [CrossRef]
14. Raschke, A.B.; Pegram, K.V.; Melkonoff, N.A.; Davis, J.; Blackwell, S.A. Collaborative conservation by botanical gardens: Unique opporutnities for local to global impacts. *J. Zool. Bot. Gard.* **2022**, *3*, 463–487. [CrossRef]
15. Wilkins, K.; Pejchar, L.; Carroll, S.L.; Jones, M.S.; Walker, S.E.; Shinbrot, X.A.; Huayhuaca, C.; Fernandez-Gimenez, M.E.; Reid, R.S. Collaborative conservation in the United States: A review of motivations, goals, and outcomes. *Biol. Conserv.* **2021**, *259*, 109165. [CrossRef]
16. Brown, J.; Puccia, L. Ex situ breeding program with wild-caught founders provides the source for collaborative effort to augment threatened New England cottontail populations. *J. Zool. Bot. Gard.* **2022**, *3*, 573–580. [CrossRef]
17. Colodner, D.; Franklin, K.; Ivanyi, C.; Wiens, J.F.; Poulin, S. Why partner with a zoo or garden? Selected lessons from seventy years of regional conservation partnerships at the Arizona-Sonora Desert Museum. *J. Zool. Bot. Gard.* **2022**, *3*, 725–737. [CrossRef]
18. Czech, B.; Krausman, P.R.; Borkhataria, R. Social construction, political power, and the allocation of benefits to endangered species. *Conserv. Biol.* **1998**, *12*, 1105–1112. [CrossRef]
19. Harris, T.R.; Heuring, W.L.; Allard, R.A.; Owens, A.K.; Hedwall, S.; Crawford, C.; Akins, C. Over 25 years of partnering to conserve Chiricahua Leopard frogs (*Rana chiricahuensis*) in Arizona, combining ex situ and in situ strategies. *J. Zool. Bot. Gard.* **2022**, *3*, 532–544. [CrossRef]
20. Montague, G. Head-starting and conservation of endangered Timber rattlesnakes (*Crotalus horridus horridus*) at Roger Williams Park Zoo. *J. Zool. Bot. Gard.* **2022**, *3*, 581–585. [CrossRef]
21. Konrad, L.; Fujii, J.A.; Hazan, S.; Johnson, A.B.; Mayer, K.A.; Murray, M.J.; Nicholson, T.E.; Staedler, M.M.; Young, C. Southern sea otter rehabilitation: Lessons and impacts from the Monterey Bay Aquarium. *J. Zool. Bot. Gard.* **2022**, *3*, 641–652. [CrossRef]
22. Maynard, L.; Cadena, B.; Thompson, T.; Pence, V.; Philpott, M.; O'Neil, M.; Pritchard, M.; Glenn, J.; Reilly, B.; Hubrich, J.; et al. Local plant and insect conservation evaluated with organizational identity theory. *J. Zool. Bot. Gard.* **2023**, *4*, 214–230. [CrossRef]
23. Sutherland, W.J.; Adams, W.M.; Aronson, R.B.; Aveling, R.; Blackburn, T.M.; Broad, S.; Ceballos, G.; Cote, I.M.; Cowling, R.M.; Da Fonseca, G.A.B.; et al. One hundred questions of importance to the conservation of global biological diversity. *Conserv. Biol.* **2009**, *23*, 557–567. [CrossRef]

24. Mallinger, M.; Markle, T.; Minerich, B.; Nordmeyer, C.; Runquist, E.; Stapleton, S. Understanding how the unique context of the Minnesota Zoo shapes our local conservation initiatives. *J. Zool. Bot. Gard.* **2023**.
25. Roznik, E.A.; Buckanoff, H.; Langston, R.W.; Smith, D. Conservation through collaboration: Regional conservation programs of the North Carolina Zoo. *J. Zool. Bot. Gard.* **2023**.

Journal of
Zoological and Botanical Gardens

MDPI

Review

Why Partner with a Zoo or Garden? Selected Lessons from Seventy Years of Regional Conservation Partnerships at the Arizona-Sonora Desert Museum

Debra Colodner *, Kim Franklin, Craig Ivanyi, John F. Wiens and Stéphane Poulin

Arizona-Sonora Desert Museum, Tucson, AZ 85743, USA
* Correspondence: dcolodner@desertmuseum.org; Tel.: +1-520-883-3021

Abstract: Zoos and botanical gardens (ZBGs) play a variety of roles in regional conservation partnerships, including their most common role as the ex situ managers of rare plant and animal populations. Using case studies from a 70-year history of conservation work at the Arizona-Sonora Desert Museum in Tucson, Arizona, USA, this paper illustrates these roles and the characteristics of ZBGs that make them versatile and effective regional conservation partners. ZBGs commonly play the role of conservation advocates, as discussed in the context of the establishment of protected islands in the Gulf of California. ZBGs also conduct field research, including the collection of long-term datasets, as exemplified by the establishment of the Ironwood Forest National Monument and a 40-year Sonoran Desert phenology database. ZBGs can be effective conveners of communities and conservation partners in regional-scale efforts, such as the Sonoran Desert Conservation Plan and Cooperative Weed Management Areas. The paper also explores the challenges faced by ZBGs in sustaining their conservation work.

Keywords: zoos; botanical gardens; local native species; regional conservation

Citation: Colodner, D.; Franklin, K.; Ivanyi, C.; Wiens, J.F.; Poulin, S. Why Partner with a Zoo or Garden? Selected Lessons from Seventy Years of Regional Conservation Partnerships at the Arizona-Sonora Desert Museum. *J. Zool. Bot. Gard.* **2022**, *3*, 725–737. https://doi.org/10.3390/jzbg3040054

Academic Editors: Gianni Bedini and Gianluigi Bachetta

Received: 8 October 2022
Accepted: 12 December 2022
Published: 19 December 2022

1. Introduction

For many people on our urbanizing planet, zoos, aquariums and botanical gardens (ZBGs) help to bridge the gap between their human-dominated environments and the rest of nature. With over 700 million annual visitors to zoos and 300 million annual visitors of botanical gardens worldwide [1,2], ZBGs help people to feel more connected to plants and other animals and learn about threats to biodiversity [3,4]. In addition to their critical role in education, ZBGs play many additional roles in the conservation arena. According to the World Association of Zoos and Aquariums, its members spend more than USD 350 million annually on field conservation and research, making them the third-largest funder of conservation worldwide [5]. ZBGs manage living collections for the ex situ conservation of rare species, host seed banks and frozen cell lines to preserve genetic diversity, conduct field research, collaborate with governments and non-governmental organizations in coordinated species and habitat conservation efforts, and engage the public in conservation activities [3,6,7]. A number of plants and animals that are or were extinct in the wild exist today due to the efforts of ZBGs [7,8].

ZBGs have a variety of management and funding structures. They are funded and managed by both public and private entities, and many depend on admission and membership fees, as well as private donations, to pay for their operations. Some (especially botanical gardens) are managed by universities or other research institutions. Due to the extensive public interactions and dependencies of most ZBGs, they also tend to be responsive to societal needs. As Christine Flanagan, former Public Programs Manager at the US Botanical Garden, explained: "More nimble than governments, more flexible than university programs, and more altruistic than private enterprise, public gardens have found ways to forge important academic, conservation and therapeutic programs" [3].

Many ZBGs whose collections and conservation efforts span the globe also participate in conservation in their own backyards, as observed from the many studies reported in this Special Issue. A smaller number of institutions focus solely on their own regions, such as the Arizona-Sonora Desert Museum (Tucson, AZ, USA), Monterey Bay Aquarium (Monterey, CA, USA), and many zoos around the world that feature local animals that cannot survive in the wild for a variety of reasons. This article uses case studies from the conservation work of the Arizona-Sonora Desert Museum to illustrate the nimbleness, flexibility, altruism, and impact of ZBGs on regional conservation efforts.

Over its 70-year history, the Arizona-Sonora Desert Museum in Tucson, AR, USA, has been a partner in numerous regional conservation efforts. The Desert Museum is a regional zoo, botanical garden, natural history museum, and art institute that immerses its visitors in the Sonoran Desert ecosystems. As a regional conservation organization, collaboration with the public agencies and communities in its region is at the core of its mission. From its earliest history, its staff have worked with regional governments to secure the protection of significant places and to hold and augment populations of rare species. The Desert Museum and its partners have learned from these experiences, sharing lessons through formal reports and publications (for example, [9–13]) and through informal professional networks.

The Desert Museum, similar to other ZBGs, has played a variety of roles in conservation partnerships. ZBGs are probably best known for their role in ex situ animal and plant care and propagation. However, education, advocacy, research, and project coordination conducted by ZBGs have made equally important contributions to regional conservation efforts (see the many projects described in this Special Issue). Government, university, and private partners turn to the Desert Museum and to other zoos and botanical gardens because of their specialized expertise, their connections with public audiences, and their versatility. ZBGs participate in both basic and applied research, engage in both in situ and ex situ conservation, work across jurisdictions, and are trusted ambassadors for conservation among many in their communities [14,15]. Although they often have more flexibility than their university or agency partners, many smaller ZBGs lack dedicated funding to support staff with a conservation focus. Nevertheless, at times, ZBGs are able to commit to long-term research, which can be difficult for university researchers tied to grants and agencies tied to annual budgets.

Here, we present several short case studies of Sonoran Desert conservation initiatives in a roughly chronological order in which the Museum played various roles, including that of conservation advocate, field researcher, public educator, project coordinator, and ex situ care and propagation manager. In each case, we focus on some of the advantages of partnering with zoos and gardens, as well as the potential challenges. The sites of these projects are shown on the map in Figure 1.

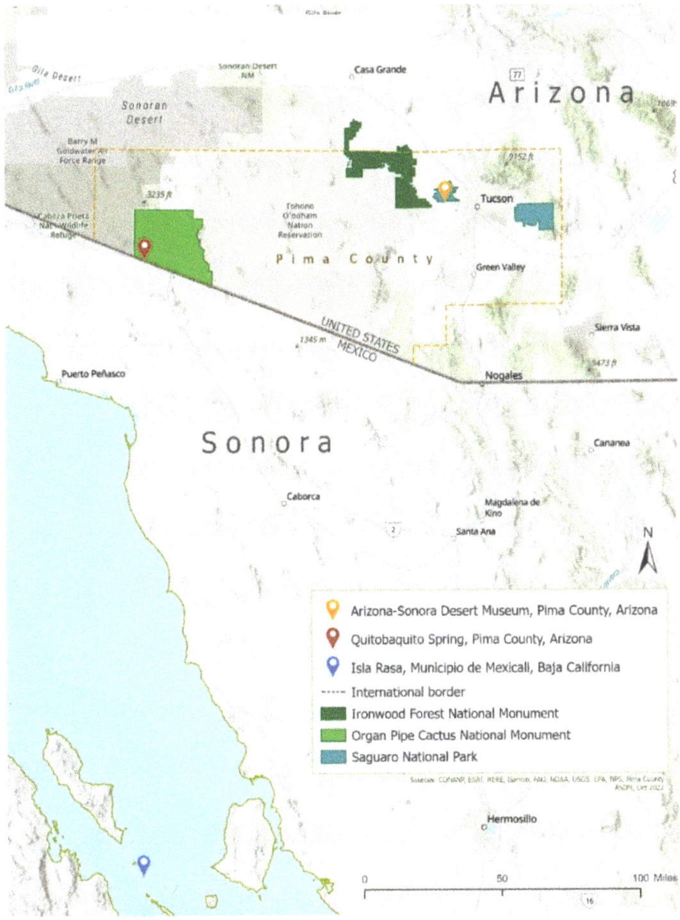

Figure 1. Regional map showing locations referenced in the text.

The Desert Museum's mission is to inspire people to live in harmony with the natural world by fostering love, appreciation, and understanding of the Sonoran Desert. The Sonoran Desert is one of the most diverse deserts in the world due to its varied geography, topography, and bimodal rainfall pattern, with summer and winter rainy seasons [16]. A variety of cultures have lived in and shaped the landscape and biodiversity for at least 13,000 years [17]. As seen in Figure 2, the Sonoran Desert region covers approximately 260,000 km^2 and includes several biotic communities, which are sometimes interlaced and alternated along the undulating basins and mountain ranges [16]. It also surrounds the extremely diverse and productive Gulf of California, with over 900 islands and islets [18]. The Sonoran Desert encompasses most of southern Arizona in the U.S., the northwestern part of Sonora, Mexico, and most of the Baja California Peninsula. It also extends slightly into southeastern California, U.S. It is home to a relatively high number of vulnerable species due to its relatively high biodiversity, habitat degradation, introduced species and pathogens, and the impacts of climate change [19,20]. Over its history, the Museum has partnered with university researchers, non-governmental organizations, indigenous communities and nations, and local, state, and federal agencies in the U.S. and Mexico to conserve the region's rich biological heritage.

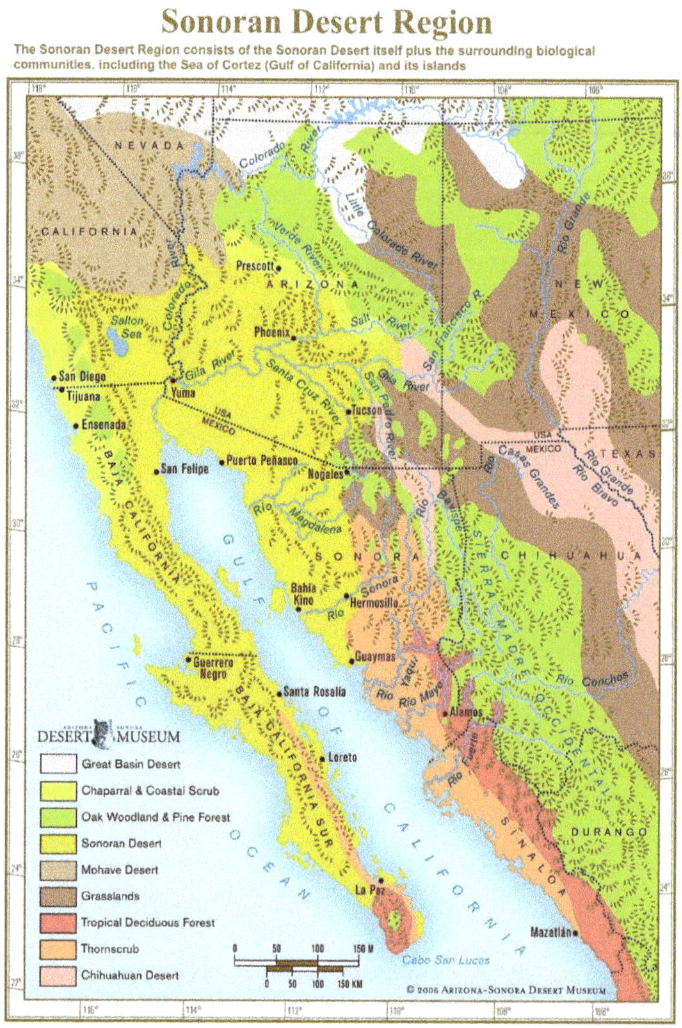

Figure 2. Sonoran Desert Region Biotic Communities.

2. Roles and Assets of Zoos and Botanical Gardens (ZBGs) in Regional Conservation Partnerships

2.1. ZBGs as Advocates: Establishment of Protection for Islands in the Gulf of California

Lewis Wayne Walker served as the associate director of the Arizona-Sonora Desert Museum from 1954 to 1970. Walker had been exploring the Baja California Peninsula and the Gulf of California since the late 1930s and first visited Isla Rasa in 1946. Located approximately 640 km south of the head of the Gulf, Isla Rasa is less than 2.6 km^2 in total and, at its highest point, not quite 30 m above sea level. The island is the primary nesting and breeding area of both Elegant Terns (*Sterna elegans*) and Heermann's Gulls (*Larus heermanni*). It is estimated that the Gulf of California islands, primarily Isla Rasa, provide a nesting habitat for 90 percent of the North American populations of these species [21]. Over subsequent visits, Walker and his colleagues noticed an astounding decline in the bird populations, mostly due to excessive egg harvesting by the residents of the Baja Peninsula. Early in the breeding season, people would destroy all the eggs they could find so that the birds would lay new eggs (without developed embryos). A few days later, they would collect the fresh eggs and sell them at local markets [22]. By the early 1960s, the bird populations had declined from more than one million nesting birds in the 1940s to approximately 25,000.

Walker published a richly illustrated article on the island in *National Geographic* in 1951 [23], which attracted additional attention in the U.S. and Mexico. He brought together influential leaders, including Dr. Joseph Wood Krutch (author, English professor, environmentalist, and pantheist), Kenneth Bechtel (founding member of the Bechtel Corporation), and William Woodin (Desert Museum Director) to revisit the islands in order to document their need for protection. He also collaborated with Dr. Bernardo Villa Ramírez from the National University of México (UNAM) and community leaders of Bahía de los Angeles on the Baja Peninsula (notably Antero Díaz). Walker enlisted the help of renowned ornithologist, Roger Tory Peterson, and the presidents of the California Academy of Sciences and National Audubon Society to publicize the importance of these habitats and the plight of its seabird populations. They worked together to produce a film, which they distributed widely to raise public awareness of the birds' plight. They also advocated in the halls of government in Mexico City and won legal protection for the island in 1964 [22]. Once established as a protected area, the island became the subject of long-term research by Mexican biologists continuing to this day.

One of the scientists visiting Isla Rasa in the 1960s was a PhD student at the Universidad Nacional Autónoma de México, Enriqueta Velarde, now posted at the Universidad Veracruzana. Dr. Velarde continued to study these seabirds, inspiring generations of Mexican scientists to further research and document the spectacular biodiversity of the Gulf of California. Their science and advocacy led to broader protections. In 1978, Isla Rasa became part of the much larger protected area under the official designation of the Área de Protección de Flora y Fauna Islas del Golfo de California, which was followed by a series of expanding protections as a Parque Nacional, a Wetland of International Importance recognized by the Ramsar Convention, and, finally, its designation as a World Biosphere Reserve in 2007 [22]. Recent surveys have documented a remarkable recovery of these birds (as well as others), with as many as 260,000 Heermann's Gulls and 200,000 Elegant Terns [22].

In the case of Isla Rasa, the public awareness and international attention made possible by the networks through which U.S. zoos operate supported the work of Mexican conservation scientists aiming to study and protect the amazing biodiversity of their country.

2.2. ZBGs in Field Research: The Ironwood Forest National Monument

Federal agencies will sometimes turn to zoos, botanical gardens, and other private conservation organizations to complete surveys on public lands that are necessary to establish or monitor management plans. In the case of the creation of the Ironwood Forest National Monument, the Desert Museum was a natural partner for the completion of the plant and animal inventories necessary to demonstrate the need for the protection of this area. Previous decades of study of the region, motivated by scientific questions arising from multiple disciplines, set the stage for the protection of this land and the Desert Museum's involvement.

Before 1970, only a few plant collections existed in this relatively undeveloped area, located just northwest of Tucson, AZ. In the 1970s and 1980s, Museum scientists began to document the flora of the area, motivated first by studies of packrat middens as records of vegetation change and later by the unusual plant communities, rare plants, and their relationships with the area's diverse geology [24]. Throughout the 1990s, the museum's botanists extended and deepened the biological surveys, discovering plants found nowhere else in the United States and mapping several rare and endangered plants for the Bureau of Land Management and U.S. Fish and Wildlife Service.

In 1998, Dr. Gary Nabhan, the museum's director of conservation and science, brought his interests and background in ethnobotany to the research of these interesting mountain ranges. He studied the ethnobotanical uses, animal interactions, and plant associations with the desert ironwood tree (*Olneya tesota*), eventually publishing a book on the topic [25], which caught the attention of local conservationists, including the county administrator. The county was engaged in a multi-decade effort to gain federal approval for a multispecies habitat conservation plan (see the Sonoran Desert Conservation Plan, below), which required the conservation of this area for rare species and for habitat connectivity. The area also contained many fragile archeological sites in need of protection. At the same time, U.S. President Clinton was eager to bolster his legacy in conservation by protecting more land as his second term was coming to an end. This parcel containing rare and unique plants and plant communities, archeological treasures, and the ecologically and culturally significant desert ironwood tree was proposed by the county for federal protection.

Given the breadth and depth of the existing museum data for this tract of land, the agencies were able to fast-track its consideration for protection, and the area was designated as the Ironwood Forest National Monument in 2000. The designation was the beginning of yet another round of study, as the Bureau of Land Management required more quantitative and comprehensive surveys of the flora and fauna of their new monument. This round of surveys also led to the discovery of a new invasive grass, buffelgrass, and motivated further study, mapping, and recommendations for its control [13].

The Monument contains 520 km^2 of varied Sonoran Desert habitat encompassing seven desert mountain ranges and protects over 600 taxa of plants and a great variety of invertebrate and vertebrate life [13]. The support of basic natural history research over the long term by the Desert Museum and the close collaboration between scientists from different disciplines (paleobiology, botany, ethnobotany, geology, and zoology) in its small science department made it a unique partner for the establishment of this protected area.

2.3. ZBGs in Long-Term Research: Phenology

Phenology, or the study of the timing of the life cycles of plants and animals, is critical to understanding basic ecological relationships, as well as the impacts of climate change on these relationships. Climate-change-induced phenological mismatches have been documented around the world in cases where pollinating insects emerge before the plants flower, birds hatch after their favorite caterpillars have peaked, or plants mature before the young animals dependent on their shoots are ready to forage [26–28]. Studying climate impacts on phenology requires extensive and long-term datasets, many of which are held by botanical gardens. Botanical gardens are also uniquely positioned to engage the public in phenological observations, as in the case of the Chicago Botanic Garden's Budburst Project [29].

The Desert Museum's horticulturalists have been collecting phenology data on dozens of native plants for nearly 40 years. Highly variable precipitation is a defining characteristic of desert regions and a confounder in phenological studies. Only now are ecologists developing sophisticated statistical techniques for studying the impact of climate change on phenology in arid regions. A recent collaboration between Conservation Science Partners, the US Geological Survey, and the Desert Museum analyzed 36 years of data on six abundant woody species and expanded on the traditional models of phenology by incorporating different metrics of moisture availability, in addition to the temperature, weather, and climate [30]. They also studied the impact of climate events on several time scales, from the daily through to the seasonal. Five of the six species studied showed advancing flowering dates in response to rising temperatures, ranging from 1 to 4 days earlier per decade, and the results for one species were uncertain. More specifically, the models suggested that flowering dates are responding to the more rapid accumulation of growing degree days occurring earlier in the year. The species' responses to moisture-related variables were less consistent and may be explained by variations in functional traits or ecological interactions, such as the mode of pollination. The study indicates the potential for phenological mismatches between pollinators and plants in this rapidly warming region of the world.

In this case, a strong connection with and desire to serve public audiences motivated the initial research. The study did not begin with climate change in mind but rather the desire to be able to answer frequently asked questions about when the peak wild flower season can be expected to occur in a given year. The scientific curiosity of the museum botanists sustained the data collection both on the museum grounds and in the adjacent Saguaro National Park. Because the data collection protocols were not exceedingly costly or time consuming, the staff and volunteers were able to sustain the data collection for 40 years. The foresight of successive museum curators and staff maintained the project despite competing priorities and a lack of funding. The value of this legacy dataset was obvious to the partners involved in the U.S. Geological Survey and the non-profit organization known as Conservation Science Partners, both of whom helped the museum with the data analysis.

2.4. ZBGs in Local Community Conservation: The Sonoran Desert Conservation Plan

Much of the desert surrounding Tucson, Pima County, AZ, is a suitable habitat for the cactus ferruginous pygmy owl (*Glaucidium brasilianum cactorum*) [31]. When the U.S. Fish and Wildlife Service (USFWS) listed it as an endangered species in 1997, local developers and environmentalists braced for years of contentious and costly court battles over new development projects. Recognizing that the owl was just one of many threatened and endangered species in this biodiversity hotspot, Pima County planners sought an alternative approach to development.

A habitat conservation plan is such an approach. When approved by the USFWS, it allows for the limited incidental loss of endangered species for development purposes when adequate mitigation plans are in place. Pima County officials began the process of developing their plan in 1998, commissioning over 200 studies and holding over 600 public meetings. They invited anyone interested to apply to serve on a project steering committee and received almost 90 applications, all of which they accepted. The Steering Committee members represented ranching, conservation, housing development, realtors, neighborhood association, transportation, mining, and other business interests. The committee members had to agree to attend nearly one year's worth of educational presentations so that they would all start from a similar place when the deliberations began [32].

In the end, this Steering Committee released a report recognizing the importance of conservation lands not only for protecting biodiversity but also for protecting the quality of life of the region's residents and its long-term economic vitality. It also recognized the need for economic growth to support the region's burgeoning population. This community-based group, representing diverse and sometimes oppositional sectors, made recommendations on the principles, funding, approaches, partnerships, duration, and many other matters regarding the Multispecies Habitat Conservation Plan for Pima County [32].

The Desert Museum served as the site for some of the Steering Committee meetings, and the museum scientists participated in the Scientific and Technical Advisory Group for the process, which created a map of the priority conservation lands, the linkages between them, and the recommendations for their management. As in the case of the Ironwood Forest National Monument, the extensive natural history and ecological information supplied by the museum was an important part of the conservation planning.

The Steering Committee report, released in 2001 [33], was an important milestone in the process of winning federal approval, which came after further studies and more detailed planning. In the years since, the museum has continued to inform the public about this community success story through exhibits, programs, and publications. The Sonoran Desert Conservation Plan, of which the Multispecies Habitat Conservation Plan is one part, covers over six million acres and helps to protect 44 vulnerable plant and animal species, as well as the ecosystem services on which the regional human population depends.

2.5. ZBGs as Ex Situ Conservation Partners

Perhaps the most common way in which ZBGs participate in local conservation partnerships is as a place for the protection and/or breeding of rare species. For example, the Association of Zoos and Aquariums reported that its member institutions spent USD 217 million on field conservation in 2021, benefitting 954 species with more than

960 partner organizations [34]. ZBGs may serve as temporary refuges during short-term habitat disruptions, work to augment populations for release or re-introduction, or they may hold long-term assurance populations of rare species. Although zookeepers and horticulturalists have expertise in managing a great variety of species, they are constantly reminded that every species is unique and that techniques which may have worked for some species may not work for others, even in the same taxon. Where ex situ conservation is being contemplated, it is important to develop partnerships with zoos and gardens early enough to develop the expertise needed to sustain the species.

Over its history, the museum has assisted in the ex situ conservation of a wide variety of plants and animals from a wide variety of taxa. As the needs for their care and breeding evolved over time, the museum found that it could be most impactful when working with species with smaller space needs, as it had a limited capacity to expand its staffing or facilities. The majority of the species of concern which it now cares for are aquatic reptiles, amphibians, and fish. Aquatic and riparian habitats have always been rare in the region and have been declining at a steady rate due to ground water pumping, surface water diversions, and a twenty-year drought [35,36]. The remaining streams and wetlands are also impacted by fires, flooding, and invasive and introduced species [35].

The Desert Museum has served as the ex situ conservation site for three species from the Quitobaquito Springs and Pond in the Organ Pipe Cactus National Monument and houses two of them currently. The Sonoyta pupfish (*Cyprinodon macularis eremus*), Sonoyta mud turtle (*Kinosternon sonoriense longifemorale*), and Quitobaquito spring snail (*Tryonia quitobaquitae*) are typical of many of the threatened and endangered aquatic species in the region [37–41]. They are endemic to a small portion of the watershed of the Rio Sonoyta and entirely dependent on heavily managed and/or impacted surface waters for their survival. The only populations of these animals in the United States live in the Quitobaquito Springs and Pond and, in the case of the snail, in two nearby springs.

Quitobaquito is a remarkable place that has been protected and managed by the generations of the Hia-Ced O'odham and Tohono O'odham, two indigenous nations of the region, as one of very few year-round sources of water. Ethnobotanist and former Desert Museum Director of Conservation and Science, Gary Nabhan, called Quitobaquito "a place in the Sonoran Desert borderlands which, more than any other I know, capsulizes what the term diversity has come to mean to both natural and social scientists alike" [19]. Additional populations of turtles and pupfish are found in small perennial stretches of the Rio Sonoyta and small ponds across the border in Sonora, Mexico. These are threatened by the continued depletion of groundwater, drought, variable seepage from sewage and dams, and the conversion of wetlands for farming and mining [41].

These aquatic species are extraordinary due to their persistence in one of the hottest, driest places in North America. The Rio Sonoyta pupfish and Quitobaquito pupfish are two subpopulations of the endangered species *Cyprinodon macularius eremus*, listed as endangered in 1986 [38]. The Sonoyta mud turtle was listed in 2017 [40]. The spring snail is a candidate for listing due to its extreme endemism [42]. The Desert Museum first became engaged in the ex situ conservation of the pupfish in the mid-1980s and of the mud turtles in 2008, aiming to provide temporary refuge to the species and establish assurance populations while hydrological studies and repairs were carried out in the area of the Quitobaquito Pond. In recent decades, the water levels of the pond have risen and fallen due to changes in precipitation and spring flow, leaks in the pond lining, and, most recently, the possible effects of groundwater pumping due to wall construction along the U.S.–Mexico border. The U.S. Fish and Wildlife Service, National Park Service, and Arizona Game and Fish Department established long-term assurance populations at the Desert Museum and Phoenix Zoo while consistently working to improve the habitat of these species.

The museum's population of pupfish survived for over three decades, although its senescence outpaced its reproduction and supplementation, and the population eventually dwindled to zero. There are currently plans to re-establish these fish at the museum. Working with the Arizona Game and Fish Department (AZGFD), the museum began turtle breeding trials in 2014. While most of the captive turtles live in separate indoor quarters, a multi-turtle outdoor habitat was designed to replicate wild conditions. After much experimentation with the habitat conditions, the first successful breeding and hatching of three turtles was accomplished in 2017, with several more baby turtles joining them over the following years.

At the request of the AZGFD and the National Park Service, the museum acquired a population of spring snails in 2022. Approximately the size of a grain of black pepper, they are extremely difficult to see, let alone count and assess. However, always ready for a new challenge, the museum keepers are learning how to keep them alive ex situ and developing care and monitoring protocols for this species.

At this time, the major conservation focus is on habitat restoration rather than captive breeding. Thus, there is no effort to breed large numbers of fish, turtles, or snails. Despite the continuing drought and fluctuations in pond water levels, the field surveys suggest that the populations in situ are currently stable. The ex situ populations are being maintained as an assurance and to continue to develop husbandry expertise as a hedge against more extreme challenges in the future. In this case, the museum's main contribution to the partnership is its expertise in, and facilities for, husbandry. The cost of this work is high (in terms of both labor and other expenses), and it is almost never fully compensated by the partnering agencies. Funding policies that prohibit labor costs are a major impediment to the support of this work. Additionally, many small institutions (including the Desert Museum) do not have federally established indirect cost rates, which limits the recovery of these costs. Although ex situ conservation work is clearly part of the mission of ZBGs, the lack of adequate funding can constrain these partnerships for smaller regional institutions.

2.6. ZBGs as Conservation Coordinators: The Regional Management of Invasive Grasses

One of the greatest threats to Sonoran Desert biodiversity is an invasive grass. Brought to the area for the purposes of soil conservation and cattle forage in the 1930s, buffelgrass (*Cenchrus ciliaris*) creates thick carpets of continuous fuel for fires in a landscape that is not adapted to fires [43]. It also outcompetes the native shrubs and cacti for water and nutrients [44]. It has the potential, especially as it is aided by climate change, to reshape one of the most diverse deserts in the world into a grassland monoculture [45]. Buffelgrass fires also threaten well-being and property in urban and rural landscapes and at the wildland–urban interface [46,47].

Invasive species, similar to many other conservation issues, are difficult to manage because they cross jurisdictions and sectors. Various land management agencies have different priorities and may have different techniques of monitoring, treatment, mapping, and documentation. There may be varying technical or policy restrictions on data sharing and treatment methods. Yet, evidently, coordinated efforts are required to make progress.

Recognizing the need for concerted action, regional conservation leaders established the Southern Arizona Buffelgrass Coordination Center in 2008. It successfully brought together dozens of landowners and managers so as to coordinate their efforts, enabling them to learn from each other and advocate together at the state and federal levels for funding. After several years of proving the concept, building a community of practitioners, developing an energetic volunteer effort, and attracting initial funding, the center's staff and advisory board realized that fluctuating funding cycles would not be able to sustain this independent non-profit endeavor in the long term. In 2016, the center approached the Desert Museum to absorb some of its functions. The Desert Museum science staff had been serving on the center's advisory board and recognized the threat posed to the Sonoran Desert. Conversation efforts with the museum's leadership, the Board of Trustees, and the Conservation and Science Team led to the decision to recognize that the threat was great enough, the museum's function in the project was clear enough, and the likelihood of external funding was good enough that this was a role which the museum could take on.

In the following years, the museum has brought together agencies, businesses, and utilities at the local, state, regional, and national levels to form the Sonoran Desert Cooperative Weed Management Area (Table 1). Through quarterly meetings, the group has shared best practices, carried out and facilitated research, pursued funding opportunities, and advocated for more coordination at the state and federal levels. The museum continues to help the group to choose and maintain a common mapping platform and address other major shared challenges, such as the inadequate workforce. The museum plays a role in research and collating and analyzing data on treatment effectiveness and alternative treatment methods. Its research has also expanded to include social scientists studying the key routes to effective partnerships in complex multi-jurisdictional conservation programs [48]. Some of their initial findings have shown the success of the first two decades of public engagement, with a strong local awareness of the status of buffelgrass as an important issue in Tucson and the surrounding Pima County.

Table 1. Partners in the Sonoran Desert Cooperative Weed Management Area.

Arizona-Sonora Desert Museum (Coordinator)
Tucson Audubon Society
Arizona Department of Transportation
Arizona Game and Fish Department
Arizona Department of Forestry and Fire Management
The Arizona Native Plant Society
Bureau of Land Management
City of Tucson
Department of Defense
Freeport McMoran
Saguaro National Park, National Park Service
Pima County
Sky Island Alliance
Southwest Vegetation Management Association
Tohono O'odham Nation
Tucson Clean and Beautiful
University of Arizona, Arizona Cooperative Extension
U.S. Fish and Wildlife Service
U.S. Forest Service
U.S. Geological Survey
Strategic Habitat Enhancements

The museum also plays an important role in public education and outreach, coordinating ongoing volunteer groups and an annual "Save our Saguaros" event. Moreover, the remarkable volunteer effort that has been effective in controlling the buffelgrass in certain areas of the Tucson Basin is a testament to the strong sense of place and shared appreciation of the Sonoran Desert ecosystem that can be at least partially attributed to the work of the museum.

Another finding derived from this work with social scientists is that a key challenge for the management of invasive species is the identification and overcoming of the barriers to collaboration and coordination between actors representing diverse institutions. These barriers include cultural traditions, mandates, and insufficient staff and funding, all of which limit the ability of institutions to look beyond their own efforts and consider the problem from a broader, more regional perspective. Identifying the barriers to and incentives for collective action will be critical for the success of efforts to address many of the ecological challenges that we will face in coming decades [48].

Over the years, partners have sometimes turned to the museum not only for its expertise but also because of a perception regarding the lower barriers to the administration of complex projects. Not only can the museum work across jurisdictional boundaries, but it also has a more flexible mandate and can help to bridge cultures. On a very practical note, small ZBGs, such as the Desert Museum, also commonly have lower overhead rates due to the lack of federally recognized rates mentioned above. This can be an advantage for project work but a disincentive for small ZBGs to take on this work.

3. Conclusions

As a regional institution, the Arizona-Sonora Desert Museum offers many informative case studies of collaboration with local and regional conservation agencies that highlight the niches that ZBGs commonly fill in these collaborations. Depending on each project's needs, ZBGs might serve as advocates for the protection of land and/or species; provide natural history data, field research, and legacy data; participate in and serve as a resource for community-based conservation efforts; provide ex situ care; and serve as coordinators or conveners of multiple partners. The nature of ZBGs as trusted translators of scientific knowledge, their ability to work across disciplines and jurisdictions, their specialized expertise in husbandry and horticulture, and their administrative nimbleness make them a valuable partner in a variety of initiatives.

Author Contributions: Conceptualization, D.C. and K.F.; writing—original draft preparation, D.C., J.W, C.I. and K.F.; writing—review and editing, D.C., K.F., C.I., S.P. and J.F.W.; project administration, D.C.; funding acquisition, C.I. All authors have read and agreed to the published version of the manuscript.

Funding: This review received no external funding.

Data Availability Statement: Not applicable.

Acknowledgments: The authors would like to acknowledge the work and contributions of the Arizona-Sonora Desert Museum conservation scientists and advocates over its 70-year history, especially the major contributors to the case studies reviewed in this article, including Richard Brusca, Mark Dimmitt, Yajaira Gray, Tani Hubbard, Gary Nabhan, and Lewis Walker.

Conflicts of Interest: The authors declare no conflict of interest.

References

1. World Association of Zoos and Aquariums. Available online: https://www.waza.org/members/waza-members/#:~{}:text=WAZA%20and%20its%20members%20are,conservation%20of%20species%20and%20nature (accessed on 24 November 2022).
2. Williams, S.J.; Jones, J.P.G.; Gibbons, J.M.; Clubbe, C. Botanic gardens can positively influence visitors' environmental attitudes. *Biodivers. Conserv.* **2015**, *24*, 1609–1620. [CrossRef]
3. Flanagan, C. The History and Significance of Public Gardens. In *Public Garden Management*; Rakow, D.A., Lee, S.A., Eds.; John Wiley: Hoboken, NJ, USA, 2011; pp. 15–29.
4. Fraser, J.; Sicker, J. *Why Zoos and Aquariums Matter Handbook*, Association of Zoos and Aquariums, Silver Spring, MD. 2008. Available online: chrome-extension://efaidnbmnnnibpcajpcglclefindmkaj/https://wzam.org/wp-content/uploads/Fraser.Sickler.2008_Why-Zoos-and-Aquariums-Matter-Handbook.pdf (accessed on 24 November 2022).
5. Mooney, A.; Conde, D.A.; Healy, K.; Buckley, Y.M. A system wide approach to managing zoo collections for visitor attendance and in situ conservation. *Nat. Commun.* **2020**, *11*, 584. [CrossRef] [PubMed]
6. Raschke, A.B.; Pegram, K.V.; Melkonoff, N.A.; Davis, J.; Blackwell, S.A. Collaborative Conservation by Botanical Gardens: Unique Opportunities for Local to Global Impacts. *J. Zool. Bot. Gard.* **2022**, *3*, 463–487. [CrossRef]
7. Minteer, B.A.; Maienschein, J.; Collins, J.P. (Eds.) *The Ark and Beyond, The Evolution of Zoo and Aquarium Conservation*; The University of Chicago Press: Chicago, IL, USA, 2018; 435p.
8. Petruzzello, M. How Botanical Gardens Save Plants with Science. Available online: https://www.britannica.com/story/how-botanical-gardens-save-plants-with-science#:~{}:text=Like%20zoos%2C%20botanical%20gardens%20often,the%20efforts%20of%20botanical%20gardens (accessed on 24 November 2022).
9. Perry-Richardson, J.; Ivanyi, C. *Captive Design for Reptiles and Amphibians, in Conservation of Endangered Species in Captivity*; Gibbons, E.F., Durrant, B.S., Demarest, J., Eds.; SUNY Press: Albany, NY, USA, 1995; Chapter 12.
10. Edwards, T. Management of captive black-tailed prairie dogs (*Cynomys ludovicianus*) utilizing natural patterns of dispersal. *Anim. Keep. Forum* **1996**, *23*, 170–177.
11. Dimmitt, M.A.; Brusca, R.C. Endangered Species and the Arizona-Sonora Desert Museum. *Endanger. Species Bull. US Fish Wildl. Serv.* **2002**, *27*, 8–12. Available online: https://digitalcommons.unl.edu/endangeredspeciesbull/6/ (accessed on 22 October 2022).
12. Bury, R.B.; Germano, D.J.; Van Devender, T.R.; Martin, B.E. Distribution, Ecology, and Conservation of Desert Tortoises in Mexico. In *The Sonoran Desert Tortoise. Natural History, Biology, and Conservation*; Van Devender, T.R., Ed.; ASDM Studies in Natural History Series; ASDM/Univ. Arizona Press: Tucson, AZ, USA, 2002; pp. 86–108.
13. Dimmitt, M.A.; Van Devender, T.R.; Wiens, J.F. Biological Survey of Ironwood Forest National Monument. Contract Report Publ. by Arizona-Sonora Desert Museum for Bureau of Land Management. 2003. Available online: https://www.desertmuseum.org/programs/ifnm_index.php (accessed on 3 November 2022).
14. Luebke, J.; Saunders, C.; Matiasek, J. Climate Change Attitudes of Zoo and Aquarium Visitors: Implications for Climate Literacy Education. *J. Geosci. Educ.* **2014**, *62*, 502–510.

15. De la Torre Dwyer, J.; Fraser, J.; Voiklis, J.; Thomas, U.G. Individual-level variability among trust criteria relevant to zoos and aquariums. *Zoo Biol.* **2020**, *2020*, 297–303. [CrossRef] [PubMed]
16. Dimmitt, M.A. Biomes and Communities of the Sonoran Desert Region. In *A Natural History of the Sonoran Desert*, 2nd ed.; Dimmitt, M.A., Comus, P., Brewer, L., Eds.; Arizona-Sonora Desert Museum Press: Tucson, AZ, USA, 2015; pp. 5–19.
17. Sheridan, T. Human Ecology of the Sonoran Desert. In *A Natural History of the Sonoran Desert*, 2nd ed.; Dimmitt, M.A., Comus, P., Brewer, L., Eds.; Arizona-Sonora Desert Museum Press: Tucson, AZ, USA, 2015; pp. 101–111.
18. Brusca, R. *Sea of Cortez in A Natural History of the Sonoran Desert*, 2nd ed.; Dimmitt, M.A., Comus, P., Brewer, L., Eds.; Arizona-Sonora Desert Museum Press: Tucson, AZ, USA, 2015; pp. 24–26.
19. Nabhan, G. Biodiversity: The Variety of Life that Sustains Our Own. In *A Natural History of the Sonoran Desert*, 2nd ed.; Dimmitt, M.A., Comus, P., Brewer, L., Eds.; Arizona-Sonora Desert Museum Press: Tucson, AZ, USA, 2015; pp. 130–135.
20. ECOS. US Fish and Wildlife Service, Listed Species Believed to or Known to Occur in Each State. Available online: https://ecos.fws.gov/ecp/report/species-listings-by-state-totals?statusCategory=Listed (accessed on 20 October 2022).
21. Velarde, E. Breeding Biology of Heermann's Gulls on Isla Rasa, Gulf of California, Mexico. *Auk* **1999**, *116*, 513–519. [CrossRef]
22. Velarde, E.; Wilder, B.; Felger, R.S.; Ezcurra, E. Floristic Diversity and Dynamics of Isla Rasa, Gulf of California—A Globally Important Seabird Island. *Bot. Sci.* **2014**, *92*, 89–101. [CrossRef]
23. Walker, L.W. The Seabirds of Isla Rasa. *Natl. Geogr.* **1951**, *99*, 239–248.
24. Wiens, J.F.; Van Devender, T.R.; Dimmitt, M.A. Vegetation and Vascular Flora of Ironwood Forest National Monument, Pima and Pinal Counties, Arizona. *Desert Plants* **2015**, *30*, 72.
25. Nabhan, G.P.; Carr, J.L. (Eds.) *Ironwood*; University of Chicago Press: Chicago, IL, USA, 1994; 92p.
26. Kudo, G.; Cooper, E.J. When Spring Ephemerals Fail to Meet Pollinators: Mechanism of phenological mismatch and its impact on plant reproduction. *Proc. R. Soc. B* **2019**, *286*, 9. [CrossRef] [PubMed]
27. Reed, T.E.; Jenouvrier, S.; Visser, M.E. Phenological Mismatch Strongly Affects Individual Fitness but not Population Demography in a Woodland Passerine. *J. Animal Ecol.* **2012**, *82*, 131–144. [CrossRef]
28. Albeck-Ripka, L.; Plumer, B. 5 Plants and Animals Utterly Confused by Climate Change. New York Times, 4 April 2018.
29. Budburst. Available online: https://www.chicagobotanical.org/research/citizen-science/budburst (accessed on 6 October 2022).
30. Zachman, L.J.; Franklin, J.F.; Wiens, K.; Crausbay, S.D.; Landau, V.A.; Munson, S. Dominant Sonoran Desert plant species have divergent phenological responses to climate change. *Madrono* **2021**, *68*, 473–486. [CrossRef]
31. Richardson, W.S.; Cartron, J.-L.; Krueper, D.J.; Turner, L.; Skinner, T.H. Chapter 3 The Status of the Cactus Ferruginous Pygmy-Owl in Arizona: Population Surveys and Habitat Assessment. In *Ecology and Conservation of the Cactus Ferruginous Pygmy-Owl in Arizona*; Gen. Tech. Rep. RMRS-GTR-43; US Department of Agriculture, Forest Service, Rocky Mountain Research Station: Ogden, UT, USA, 2000; pp. 27–46.
32. Sonoran Desert Conservation Plan. Available online: https://webcms.pima.gov/government/sustainability_and_conservation/conservation_science/the_sonoran_desert_conservation_plan/ (accessed on 6 October 2022).
33. Sonoran Desert Conservation Plan Steering Committee. Available online: Chrome-extension://efaidnbmnnnibpcajpcglclefindmkaj/https://webcms.pima.gov/UserFiles/Servers/Server_6/File/Government/Office%20of%20Sustainability%20and%20Conservation/Conservation%20Sciece/sdcp%20reports/SDCP-Steering-Committee-Report-to-the-Pima-County (accessed on 6 October 2022).
34. Association of Zoos and Aquariums, Annual Report on Conservation and Science. 2021. Available online: Chrome-extension://efaidnbmnnnibpcajpcglclefindmkaj/https://assets.speakcdn.com/assets/2332/aza_arcshighlights_2021_final_web.pdf (accessed on 22 October 2022).
35. Zaimes, G. (Ed.) Understanding Arizona's Riparian Areas, Arizona Cooperative Extension, AZ 1432. 2007. Available online: Chrome-extension://efaidnbmnnnibpcajpcglclefindmkaj/https://extension.arizona.edu/sites/extension.arizona.edu/files/pubs/az1432.pdf (accessed on 22 October 2022).
36. Archer, S.; Predick, K.I. Climate Change and Ecosystems of the Southwestern United States. *Rangelands* **2008**, *30*, 23–28. Available online: https://bioone.org/journals/Rangelands/volume-30/issue-3 (accessed on 22 October 2022).
37. Miller, R.R.; Fuiman, L.A. Description and Conservation Status of *Cyprinodon macularis eremus*, A New Subspecies of Pupfish from Organ Pipe Cactus National Monument, Arizona. *Copeia* **1987**, *3*, 593–609. [CrossRef]
38. ECOS. US Fish and Wildlife Service, Desert Pupfish. Available online: https://ecos.fws.gov/ecp/species/7003 (accessed on 20 October 2022).
39. Conner, C.W. The Quitobaquito Desert Pupfish, an Endangered Species within Organ Pipe Cactus National Monument: Historical Significance and Management Challenges. *Nat. Resour. J.* **2000**, *40*, 379–410.
40. NatureServe Explorer. Sonoyta Mud Turtle. Available online: https://explorer.natureserve.org/Taxon/ELEMENT_GLOBAL.2.105200/Kinosternon_sonoriense_longifemorale (accessed on 22 October 2022).
41. Knowles, G. Aquatic life in the Sonoran Desert. *Endanger. Species Update* **2003**, *20*, 22. Available online: link.gale.com/apps/doc/A114168551/AONE?u=uarizona_main&sid=bookmark-AONE&xid=d71408b0 (accessed on 2 November 2022).
42. NaureServe Explorer. Quitobaquito Tryonia. Available online: https://explorer.natureserve.org/Taxon/ELEMENT_GLOBAL.2.117781/Tryonia_quitobaquitae (accessed on 3 November 2022).
43. McDonald, C.J.; McPherson, G.R. Fire behavior characteristics of buffelgrass-fueled fires and native plant community composition in invaded patches. *J. Arid. Environ.* **2011**, *75*, 1147–1154. [CrossRef]

44. Franklin, K.A.; Lyons, K.; Nagler, P.L.; Lampkin, D.; Glenn, E.P.; Molina-Freaner, F.; Markow, T.; Huete, A. Buffelgrass (*Pennisetum ciliare*) land conversion and productivity in the plains of Sonora, Mexico. *Biol. Conserv.* **2006**, *127*, 62–71. [CrossRef]
45. Franklin, K.A.; Molina-Freaner, F. Consequences of Buffelgrass Pasture Development for Primary Productivity, Perennial Plant Richness, and Vegetation Structure in the Drylands of Sonora, Mexico. *Conserv. Biol.* **2010**, *24*, 1664–1673. [CrossRef] [PubMed]
46. Wilder, B.T.; Jarnevich, C.S.; Baldwin, E.; Black, J.S.; Franklin, K.A.; Grissom, P.; Hovanes, K.A.; Olsson, A.; Malusa, J.; Kibria, A.S.; et al. Grassification and Fast-Evolving Fire Connectivity and Risk in the Sonoran Desert, United States. *Front. Ecol. Evol.* **2021**, *9*, 5561. [CrossRef]
47. Brenner, J.C.; Franklin, K.A. Living on the edge: Emerging environmental hazards on the peri-urban fringe. *Environ. Sci. Policy Sustain. Dev.* **2017**, *59*, 16–29. [CrossRef]
48. Lien, A.M.; Baldwin, E.; Franklin, K. Collective Action and Invasive Species Governance in Southern Arizona. *Rangel. Ecol. Manag.* **2021**, *74*, 151–164. [CrossRef]

Journal of
Zoological and Botanical Gardens

MDPI

Review

Understanding How the Unique Context of the Minnesota Zoo Shapes Our Local Conservation Initiatives

Mary Mallinger [1,*], Tricia Markle [1], Ben Minerich [1], Cale Nordmeyer [1], Erik Runquist [1] and Seth Stapleton [1,2]

1 Department of Conservation and Research, Minnesota Zoo, Apple Valley, MN 55124, USA
2 Department of Fisheries, Wildlife and Conservation Biology, University of Minnesota, Saint Paul, MN 55455, USA
* Correspondence: mary.mallinger@state.mn.us

Abstract: The field of wildlife conservation is comprised of a variety of players with different contexts and approaches. Zoos and aquariums, historically largely focused on public entertainment, are shifting more towards conservation-minded missions and can play a unique role in wildlife conservation by leveraging their distinct assets. The Minnesota Zoo is an AZA-accredited institution and an agency of the State of Minnesota that has been conducting wildlife conservation for over 40 years. Here, we review our current portfolio of local field projects, including initiatives targeting pollinators, native mussels, turtles, and bison, using several considerations to structure and better understand how our unique context has shaped our work. Our designation as a state agency has impacted our initiatives by necessitating a focus on local efforts and has facilitated many partnerships with other government agencies. Indeed, partnerships have been vital to our success and have shaped our programs significantly since their inception. All of the Zoo's conservation initiatives are built on a bedrock of sound science, and we continue to contribute to the field through research, utilizing the expertise of department staff. In addition, the various funding streams that support our programs have dramatically shaped our work and have created some siloing of staff within the department. However, grant funding can serve as a buffer against the impacts of economic uncertainty, as evidenced during the COVID-19 pandemic. Lastly, our programs have expanded and our objectives have pivoted over the years in response to changing needs and opportunities; such flexibility—and increased flexibility for our staff—is imperative to the future success of these efforts. The Minnesota Zoo's narrative is unique and helps us understand how we can continue to most effectively carry out local conservation efforts. As we work to protect habitats and save species from extinction, it is important to utilize the distinct assets that each organization can contribute in order to have the greatest collective impact.

Keywords: bison; butterfly; ex situ conservation; government agency; in situ conservation; native mussels; turtles; zoos and aquariums

Citation: Mallinger, M.; Markle, T.; Minerich, B.; Nordmeyer, C.; Runquist, E.; Stapleton, S. Understanding How the Unique Context of the Minnesota Zoo Shapes Our Local Conservation Initiatives. *J. Zool. Bot. Gard.* **2023**, *4*, 427–444. https://doi.org/10.3390/jzbg4020032

Academic Editors: Ursula Bechert and Debra Colodner

Received: 17 November 2022
Revised: 1 May 2023
Accepted: 10 May 2023
Published: 17 May 2023

1. Introduction

1.1. A Collaborative Approach to Conservation

The practice of conserving wildlife and their habitats is complex and multi-faceted. There are many ways to analyze and categorize the diverse approaches to wildlife conservation [1], but broadly, wildlife conservation practices may be separated into two categories—ex situ and in situ conservation [2]. Traditionally, wildlife biologists, researchers, land managers, and conservationists work together to develop and implement conservation practices for wildlife and their habitats in the field (i.e., in situ). Meanwhile, ex situ efforts focus on the conservation and maintenance of animal populations outside their natural habitats, often through breeding and rearing programs. In recent years, however, there has been increased recognition of the value of a more integrated approach to wildlife conservation wherein these two tracks function together [3]. The Conservation Planning

Specialist Group's (CPSG) One Plan approach embodies this integrated process and is modeled on "the development of management strategies and conservation actions by all responsible parties for all populations of a species, whether inside or outside their natural range" [4]. This requires that conservation planning for a species simultaneously considers the management of both captive and wild populations.

1.2. Role for Zoos and Aquariums

The science and practice of wildlife conservation have been orchestrated by a variety of organizations for as long as the field has formally existed. Non-profit organizations, government agencies from the federal level to local community-based organizations, and educational institutions such as universities are all actively involved in conservation efforts today, each filling its own niche within the broader conservation landscape. Zoos and aquariums, often operating as not-for-profit and public organizations [5], primarily have focused on ex situ conservation but have supported in situ initiatives as well. These ex situ efforts have influenced the field through coordinated breeding programs that can create "insurance" populations for at-risk animals and may serve to augment wild populations. While captive breeding programs are coordinated almost entirely outside of a species' natural range, these managed populations may provide a source for population augmentation, such as reintroducing species to their wild habitats. A well-known example is the Asian wild horse (Przewalski's horse; *Equus ferus przewalskii*). Once found throughout the grassland steppes of Europe and Asia, the Asian wild horse was officially declared extinct in the wild in the 1960s [6]. After years of ex situ breeding efforts coordinated largely by zoos, these animals were reintroduced in the wild in partnership with local organizations, leading to the successful establishment of in situ populations in areas of Europe and Asia [7].

Zoos and aquariums are uniquely positioned to support a collaborative One Plan approach to wildlife conservation. Institutions typically maintain large and diverse collections of animals that may be managed to meet ex situ conservation objectives (e.g., as a source for conservation translocations) and have the ability to reach large audiences with conservation messaging and encourage public action. They often participate in captive breeding programs and thus maintain staff and institutional expertise in reproductive biology and small population management. Zoos and aquariums can conduct important trial research to help create and inform conservation protocols when pre-existing knowledge is lacking, such as the development of breeding and rearing techniques for species that have not been managed in a captive setting previously. In addition, zoos and aquariums are very specialized facilities with spaces and equipment designed specifically for the rearing and caring of animals. Staff are highly experienced in animal husbandry which can be a significant asset for head-starting and reintroduction programs. Zoo veterinary expertise also helps manage health-related concerns of head-starting and insurance populations and ensures that animals designated for translocations are in adequate physical condition.

The Association of Zoos and Aquariums (AZA) is a non-profit organization "dedicated to the advancement of zoos and aquariums in the areas of conservation, education, science, and recreation" [8]. There are currently 238 institutions across 13 countries that are accredited by AZA, collectively drawing more than 200 million visitors annually [9]. About 900 species housed at AZA facilities are classified by the International Union for Conservation of Nature (IUCN) as Vulnerable, Endangered, Critically Endangered, or Extinct in the wild [5]. AZA institutions don't just house animals from across the globe, however. AZA institutions are also actively and increasingly involved in field conservation efforts, with 117 recognized reintroduction programs coordinated by AZA members [5]. In 2020, AZA institutions collectively spent >$208 million on field conservation [9].

1.3. Minnesota Zoo

The Minnesota Zoological Garden (MN Zoo) is an AZA-accredited institution located in Apple Valley, Minnesota, USA. Situated on nearly 500 acres, the MN Zoo houses close to

5000 animals and over 490 species, approximately 15% of which are listed as imperiled at international, federal, or state levels. As the largest environmental education institution in the state, we welcome over 1.3 million guests through our doors in a typical year (before the COVID-19 global pandemic). We are one of only two zoos in the United States that operates as a state agency (the North Carolina Zoo is an agency of the North Carolina Department of Natural and Cultural Resources), which makes our institution relatively unique and impacts our conservation approach in a variety of ways.

Our current conservation portfolio includes both local and global species. Most of our work, however, takes place in Minnesota with efforts focusing on the conservation of pollinators, freshwater turtles, native freshwater mussels, and North American plains bison (*Bison bison*). The origin of these programs was significantly shaped by our unique circumstances as both an AZA-accredited zoo and aquarium and a state agency, and this impact continues today. Here, we review our portfolio of local conservation projects and identify common threads among those programs that allow us to understand how we, as a zoo and an agency of the state of Minnesota, are uniquely positioned to contribute to wildlife conservation in the Upper Midwest USA.

2. Case Studies

The MN Zoo's active local projects address the conservation of pollinators, freshwater turtles, native mussels, and North American plains bison. We evaluate each program using four guiding parameters to lend structure to the review and maintain consistency. We begin by providing the reader with programmatic context, including the target species, threats, local conservation status, and initial programmatic objectives and activities. Next, we consider primary partnerships that have influenced the program, including historical, current, and developing collaborations. The overall growth of each initiative throughout the years is examined, followed by the main source(s) of funding.

After reviewing each of our Minnesota-based conservation initiatives using these parameters, we identify common themes and detail lessons learned throughout the remainder of the paper.

2.1. Pollinator Conservation Initiative

2.1.1. Context

The MN Zoo's Pollinator Conservation Initiative (PCI) began in 2012 with the intent to focus on field surveys and research, and the possible eventual reintroduction of two of the state's vanishing butterfly species. Minnesota once boasted about 18 million acres of rolling tallgrass prairie [10] which supported a diverse array of wildlife. Over the last century, the vast majority of that landscape was converted to agriculture and development [11] and many of the species that relied on the prairie were lost with it [12–14]. The Dakota skipper (*Hesperia dacotae*) and Poweshiek skipperling (*Oarisma poweshiek*) were both commonly found throughout Minnesota's prairies [15,16] but have experienced dramatic population declines range-wide in recent decades [17,18]. Both species are now listed under the U.S. Endangered Species Act [19]; the Dakota skipper is categorized as Endangered [20] and the Poweshiek skipperling as Critically Endangered [21] by the IUCN. They are among a handful of butterflies listed at the State level by the Minnesota Department of Natural Resources (DNR).

2.1.2. Partnerships

When the PCI first started, little was known about the state of wild populations of Dakota skipper and Poweshiek skipperling. The MN Zoo's Conservation staff were in close and regular contact with other government agencies including the DNR and the United States Fish and Wildlife Service (USFWS). Biologists within these agencies expressed concern that these species were declining as reported sightings became less frequent and both were undergoing listing assessments by USFWS. Our initial role was to survey sites where the species were known to historically occur in the hopes of gaining a better understanding of their population status. As surveys continued over the first few years, it became clear that Dakota skipper populations had declined rapidly and that the Poweshiek skipperling was on the verge of extirpation in—if not already extirpated from—Minnesota [19].

Through a series of subsequent workshops (including a variety of partners) following ex situ planning guidelines established by the IUCN [22], potential roles for ex situ conservation programs were identified to support the recovery of both butterfly species in North America [23]. The MN Zoo was uniquely positioned to aid in these specific efforts. As an institution with nearly 40 years of experience rearing and breeding animals for conservation initiatives, we were able to lend our expertise and build the ex situ aspect of the butterfly program. Although ex situ breeding and reintroduction programs have been an increasingly implemented conservation tool [24,25], initially no protocols existed for rearing and breeding these species or any grass skipper (sub-family Hesperiinae). The MN Zoo's biologists worked closely with other MN Zoo staff to design appropriate facilities (Figure 1) and began to build the protocols necessary for the rearing and managed breeding of Dakota skipper and Poweshiek skipperling.

Figure 1. Imperiled Dakota skipper larvae are reared on host plants inside butterfly hoop houses at the Minnesota Zoo. Poweshiek skipperling are reared in similar facilities on campus.

2.1.3. Growth

Once husbandry protocols were successfully developed, the PCI expanded quickly. Adhering to the IUCN's Translocation guidelines [26], we have partnered closely with The Nature Conservancy and DNR on Dakota skipper conservation, as well as with collaborators in Michigan, where the last remaining wild populations of Poweshiek skipperling in the USA are now found, to coordinate the site surveys and determine the scope and form of releases of zoo-bred butterflies. Both Dakota skipper and Poweshiek skipperling are prairie obligate species, and candidate sites for reintroduction and augmentation are located hundreds of miles from the MN Zoo campus. As a zoo, the management of natural areas and sites lies outside our purview. Thus, partnerships have been critical to the implementation of in situ conservation work as well as the overall growth and success of this program. The PCI began with efforts focused on head-starting to help augment wild

Figure 2. A reintroduced male Dakota skipper reared at the Minnesota Zoo perches atop a purple coneflower at a reintroduction site in southwestern Minnesota.

populations and has since shifted to managing captive breeding populations which serve as a source for translocations (Figure 2). We are now able to rear about 3000 Dakota skipper and approximately 400 Poweshiek skipperling per year for breeding and release efforts. The list of collaborators is also expanding. We are currently working with biologists at John Ball Zoo and Assiniboine Park Zoo which has doubled the capacity to rear and translocate Poweshiek skipperling. Additional accredited zoos may be joining Dakota skipper ex situ efforts in the near future.

The PCI has continued to shift and grow over the last decade as needs have been identified and research has developed. Closer to home, the PCI has recently expanded into research and outreach for other pollinators, particularly the U.S. Endangered rusty patched bumble bee (*Bombus affinis*) which exists wild on the MN Zoo campus. MN Zoo biologists are involved in federal recovery planning efforts for all these species, utilizing quantitative decision tools and facilitating large partnerships. Additional research conducted by MN Zoo biologists has included studies on some of the probable stressors on globally endangered Poweshiek skipperling and Dakota skipper. For example, we have documented the near-universal presence over many years of multiple agricultural pesticides, particularly broad-spectrum insecticides targeting the economically damaging soybean aphid, within federally designated Critical Habitat for these protected butterflies in the Upper Midwest. We have also shown experimentally that some invasive grass species in these same prairies are likely presenting ecological sinks to Dakota skippers by decreasing survivorship and delaying the development of their larvae that may feed on these less suitable hosts [27].

All our field conservation programs have sets of long-term goals, objectives, and associated actions to structure our work. Objectives and activities are established in coordination with our partners and, where possible, guided by formal recovery plans, such as with Poweshiek skipperling work. Our progress toward meeting objectives is revisited regularly and we adapt our work as necessary to address emergent needs and unforeseen issues. Success is evaluated based on a variety of metrics that are specific to each program.

For PCI work, metrics include survival at key life history stages, breeding rates of captive-reared animals, total numbers of individuals released via population augmentation efforts, and hatching rates of wild-collected eggs. We also evaluate success based on our ability to disseminate key findings and activities to the scientific community via publications in the peer-reviewed literature, presentations at professional conferences and meetings, and technical reports.

2.1.4. Funding

As is the case with much conservation work, external funding has played a significant role in shaping this initiative. Minnesota's Environment and Natural Resources Trust Fund (ENRTF) is awarded annually to a variety of grantees "for the public purpose of protection, conservation, preservation, and enhancement of the state's air, water, land, fish, wildlife, and other natural resources" [28]. The MN Zoo's Dakota skipper operations have been funded in large part by the ENRTF over the last decade which mandates that appropriations be awarded to projects operating within the state. Thus, the reintroductions of Dakota skipper to date have necessarily been constrained to Minnesota. As the remaining known sites for Poweshiek skipperling are outside Minnesota we cannot use State funds for this project. The PCI has largely depended on federal funding, via grants such as the Great Lakes Restoration Initiative (GLRI), to execute Poweshiek skipperling work. Additional programmatic support is provided by the MN Zoo, Minnesota Zoo Foundation, and other organizations. The ENRTF and GLRI funds have provided primary support for two dedicated full-time entomologists who focus exclusively on these programs, as well as a few seasonal staff during peak husbandry and field season.

2.2. Freshwater Turtles

2.2.1. Context

Minnesota, also known as the 'Land of 10,000 Lakes', is home to an abundance of freshwater resources. Nine species of turtles are native to the state including the wood turtle (*Glyptemys insculpta*), Minnesota's most terrestrial turtle species [29]. Habitat degradation and destruction, collection for the pet trade, and increased road mortality and predation [30], compounded by the species' low reproductive potential, have impacted the population throughout its range in North America [31,32]. In 1984, the wood turtle was listed as threatened by the DNR and is now only found at a few sites within the state [33]. The USFWS is completing a status assessment for potential listing under the Endangered Species Act on both the wood turtle and Blanding's turtle (*Emydoidea blandingii*), another species native to Minnesota [34,35].

2.2.2. Partnerships

Similar to the Pollinator Conservation Initiative, the MN Zoo's freshwater turtle conservation program began in partnership with other government agencies, including the DNR and the Minnesota Department of Transportation (MnDOT). Because both wood and Blanding's turtles are protected in Minnesota, the DNR has been conducting research and implementing conservation activities for years [29,36] and our objective was to support their ongoing research and conservation efforts. Specifically, in 2017, we began partnering with the DNR to use GPS telemetry to document the movements and habitat use of wood turtles and to identify and protect key nesting sites, and with MnDOT to mitigate road mortality.

2.2.3. Growth

Over the first few seasons, we collected valuable information on the spatial ecology of wood turtles including range size, habitat use, and the distribution of nesting sites. However, it also became evident that wood turtles were suffering from poor recruitment due to factors including high rates of nest predation such that additional measures would be needed to maintain local populations while threats could be addressed. As such, with the support of the DNR we initiated a formal head-starting program whereby eggs would be collected from the wild and hatchlings reared through their first year of life before being released to natal sites to help bolster wild populations. The head-starting program has grown considerably over the years and program staff at the MN Zoo now rear approximately 30 turtles from hatching through their first year (Figure 3). Before release back into their natural habitats, a sample of the yearling wood turtles is outfitted with VHF transmitters so they can be routinely tracked by biologists. Little is currently known about the habitat use and survivorship of yearling wood turtles [29], and these data will inform the conservation of the species and help to evaluate the success of the head-starting program.

Figure 3. Wood turtle hatchlings are reared at the Minnesota Zoo for their first year of life as part of the head-starting program.

In addition, MN Zoo biologists partnered with MnDOT to study road mortality and mitigation strategies via a Before-After Control-Impact study design [37]. This research suggested that standard chain link fences outfitted with j-hook wrap-around end treatments can be an effective means to mitigate the mortality of turtles on Minnesota's roadways (Figure 4). At some sites, turtle mortality was reduced by 90% with the installation of fencing [37]. While the initial research project concluded in 2022, MN Zoo biologists continue to work with staff and researchers at MnDOT and other local road authorities to implement and improve roadside fencing in order to reduce mortality and ensure turtles and other wildlife have safe passageways.

Lastly, we are expanding our partnership with the DNR and delving into a new initiative within the freshwater turtle program. While widespread across the state, Blanding's turtles have experienced range-wide population declines over the last decades [38]. The southeastern corner of Minnesota hosts one of the largest historical populations of Blanding's turtles in North America [39], but little is known about the current status of this population. MN Zoo and DNR biologists are conducting pilot work to inform a larger-scale study that will estimate demographic metrics, assess the status and evaluate the resiliency of this important population in Minnesota.

Figure 4. Chain-linked fences with wrap-around end treatments and retrofitted with fine mesh hardware cloth at the base effectively reduce the mortality of turtles on Minnesota's roads.

Key performance indicators for our turtle conservation program have varied by initiative (i.e., head-starting and documenting spatial ecology of wood turtles, mitigating road mortality, and updating the demographic status of Blanding's turtles). Specific metrics have included the number of transmitters deployed on wood turtles, number of eggs collected from the wild, hatching rates of wild-collected eggs, growth rates of animals in our care, number of individuals released via head-starting efforts, post-release survival of head-started wood turtles, percent reductions in mortality after implementation of mitigation measures, and trap nights and capture rates of Blanding's turtles (which will provide the foundation for our demographic assessments).

2.2.4. Funding

Primary financial support for our freshwater turtle program is provided by Minnesota's ENRTF, which has awarded grants to the MN Zoo for these efforts since 2018, and a contract with MnDOT which supported our research on the efficacy of small animal exclusion fencing in reducing the mortality of turtles and other wildlife. In addition, we receive supplemental funding from the MN Zoo's state appropriation and the Minnesota Zoo Foundation. Staffing currently includes one full-time biologist who exclusively implements turtle research and conservation activities and a seasonal internship that is shared with the freshwater mussel program.

2.3. Native Freshwater Mussels

2.3.1. Context

Freshwater mussels are found throughout much of the world, but North America is a biodiversity hotspot, boasting nearly 300 species of native mussels [40]. Unfortunately, native mussel populations in the United States have experienced steep declines since the late 1800s [41]. Historic overharvesting of mussels for the pearl button industry, the construction of dams, pollution, and aquatic invasive species are among the factors that have led to the categorization of freshwater mussels as the most at-risk group of animals in North America today [42].

Much of Minnesota is dotted with freshwater lakes and wetlands, with miles of winding river carving through the landscape, providing plentiful potential habitat for mussels. The state is home to dozens of species of native freshwater mussels, which play a crucial role in the health and vitality of Minnesota's aquatic resources [43]. The DNR has studied and surveyed native mussels since the late 1990s and propagates and conducts reintroductions for some of the 28 species of conservation concern at the State and federal levels [44].

2.3.2. Partnerships

The MN Zoo began its involvement in mussel conservation in 2015 by partnering with and supporting the efforts of the DNR. Following consultation with the DNR, we concluded that we could best contribute via head-starting juvenile stock provided by partner organizations. With years of experience rearing species with similar life histories and requirements as mussels, we were well-positioned as a partner for ex situ conservation work. MN Zoo staff expertise with coral propagation and the maintenance of aquatic systems has been an asset to building the infrastructure for the freshwater mussel program. The MN Zoo is also located on hundreds of acres of land, much of it undeveloped, with access to several small lakes and wetlands that have been critical to the expansion of the ex situ efforts.

2.3.3. Growth

The mussel program began at the MN Zoo by rearing a few hundred juvenile mussels in five-gallon buckets in one of the small lakes on our grounds (Figure 5). Our work and capacity to rear mussels have expanded significantly since the program's inception in 2015. We constructed docks at an alternative lake with high-quality water and installed new baskets for rearing. In collaboration with staff from the Aquariums and Life Support departments, we designed and constructed a new rearing and research facility that further expanded our physical capacity (Figure 6). The addition of the rearing facilities also allowed us to construct specialized systems for newly transformed mussels, which are nearly microscopic in size and notoriously difficult to rear. Today, the MN Zoo's mussel-rearing facilities collectively have the capacity to house >100,000 individuals.

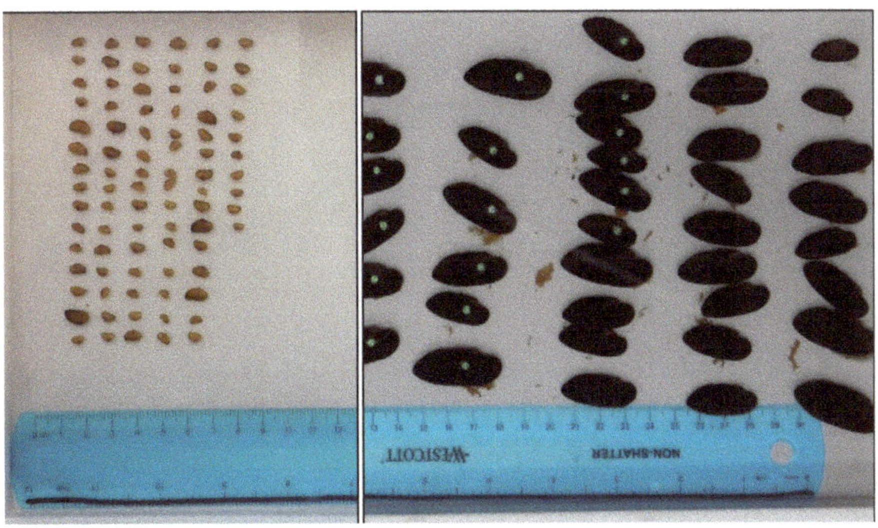

Figure 5. The Minnesota Zoo head-starts native mussels to support recovery efforts led by the Minnesota Department of Natural Resources. Mussel growth is demonstrated by comparing individuals at one year of age (**left**) and after one year of head-starting at the MN Zoo (**right**).

Figure 6. The Minnesota Zoo's native mussel facility includes individual rearing pans which allow for manipulation of metrics such as substrate, flow rate, and mussel density to facilitate experimentation. Water is circulated from a nearby lake.

The propagation of mussels has been studied for years, but much remains unknown about the natural history of native mussels and how best to rear many species [45]. As such, in addition to helping expand the rearing capacity of our partners, we have conducted research to inform these husbandry protocols and systems. Current studies include how the presence of larger size classes and different substrates impact the growth and survival of individuals.

We are now supporting the conservation of native mussels in other new ways. Freshwater mussels require a fish host to complete their development, and DNR researchers rear host fish as part of their propagation programs. Establishing and maintaining the appropriate facilities for the husbandry of both freshwater mussels and their fish host species requires significant space, time, and experience. The MN Zoo recently began rearing young walleye (*Sander vitreus*), a common mussel host fish, provided as fingerlings from local hatcheries to aid in the successful propagation of mussel species. The pilot year was successful with high survival, and we subsequently built permanent structures so that we can continue rearing ~200 walleye a year to support mussel conservation efforts. Measures of success for our mussel program include survivorship and growth of animals in our care and numbers of animals returned to the DNR for eventual release into the wild (or with fish hosts, for inoculation with mussel larvae) as a result of our head-starting programs.

We are also currently conducting research in collaboration with the University of Minnesota's Aquatic Invasive Species Research Center (MAISRC) on the invasive zebra mussel (*Dreissena polymorpha*). Originally native to freshwater in Eurasia, zebra mussels likely arrived in the U.S. in the 1980s and have negatively impacted our lakes and rivers and native mussel populations dramatically [46]. In 2020, MN Zoo biologists began conducting research with scientists at MAISRC on how to rear zebra mussels in a controlled setting. The goal is to establish rearing protocols and to provide zebra mussels at all life stages for researchers working on eradication and control methods.

2.3.4. Funding

As with our other programs, Minnesota's ENRTF has played a critical role in the establishment and growth of the mussel conservation program. Initial funding from the ENRTF provided primary support for staff time and the construction of our mussel research and rearing facility. A subsequent award from ENRTF has supported the maintenance and further expansion of our work; supplemental funding is provided by the MN Zoo state appropriation and the Minnesota Zoo Foundation. The MN Zoo's mussel conservation program is staffed by one full-time biologist and a seasonal internship shared with the turtle program, with additional support provided by the Life Support and Aquariums departments. The zebra mussel research is currently funded by the University of Minnesota via the Minnesota Aquatic Invasive Species Research Center.

2.4. North American Plains Bison

2.4.1. Context

North American bison once numbered in the tens of millions and inhabited great expanses of the continent [47]. Throughout the 19th century, bison were hunted to near extinction by European settlers until fewer than 1000 individuals remained [48]. Coordinated conservation efforts helped save the species from extinction and resulted in successful reintroductions at sites across North America [49]. During the recovery period, however, domestic cattle (*Bos taurus*) were allowed to breed with many of the few remaining bison populations, altering their genomic makeup [48]. Previous studies have estimated that approximately one percent of today's North American bison are free of cattle genes [48], although more recent evidence suggests that all bison today have some degree of cattle genomic introgression [50]. The overall resulting effects of cattle genes within bison populations, an animal that has existed in North America for thousands of years, are largely unknown [51].

For the long-term conservation of the species, it is critical that bison genetics are managed through healthy herds across their historical range. Many of the original populations reintroduced on public lands over the last few decades, such as at Yellowstone National Park, have reached carrying capacity. The National Park Service recommended that smaller satellite herds, such as those at state parks, be established and managed to ensure the future health and conservation of bison in North America [52].

2.4.2. Partnerships

In 2012, the Parks & Trails Division of the DNR and the MN Zoo joined together to form the Minnesota Bison Conservation Herd (MBCH). The partnership was established with two primary goals: to reestablish a Minnesota bison population of sufficient size to sustain genetic diversity and contribute to bison conservation in North America, and to provide increased opportunities for visitors to connect with the species. The targeted population of the MBCH is set at 500 individuals, made up of several smaller satellite populations but managed collectively [52]. The well-established collaborative relationship between the MN Zoo and the DNR has aided the conservation efforts substantially. The DNR has jurisdiction over many acres of land that could provide potential bison habitat. Because of the unique history of bison in North America and their breeding with cattle, it is important to understand and actively manage the genetic makeup of the overall herd. MN Zoo staff have extensive experience in cooperative breeding and studbook management and have been able to leverage this expertise in support of the bison program. In order to reach the targeted goal of 500 individuals and to ensure the future success of the species, partnerships have been and will continue to be essential.

2.4.3. Growth

The Parks & Trails Division of the DNR has been managing bison in the state since three animals were reintroduced to Blue Mounds State Park in 1961. Since then, the herd at Blue Mounds State Park has grown substantially (Figure 7), and a new herd was reintroduced at Minneopa State Park in 2015. Expansion of the MBCH has continued with the addition of new partners and the creation of more satellite herds. Today, the MBCH is comprised of approximately 130 animals at five sites across the state: two state parks, two zoos (including the MN Zoo), and a county park. There remains considerable opportunity for growth as additional organizations have recently expressed interest in establishing or expanding bison herds as a part of the MBCH.

Figure 7. Bison at Blue Mounds State Park, part of the Minnesota Bison Conservation Herd, are monitored and provided health checks during the annual management event in October 2022.

We measure the success of this program via metrics such as the overall genetic diversity of the MBCH, minimization of cattle genomic introgression, the size of the MBCH herd (relative to carrying capacities of individual sites and our stated target of 500 individuals), and the expansion of the MBCH to incorporate new partner organizations from across the State.

2.4.4. Funding

Funding and support for the bison conservation work come from a variety of sources. The MN Zoo's Animal Care staff, who care for the MN Zoo's bison herd and support the annual management event, and the Animal Health staff, who provide healthcare to our bison and some MBCH satellite herds, are primarily supported through our general operating budget. Individual and institutional contributions made to the Minnesota Zoo Foundation help fund the genetic analysis, conducted by the DNA Technologies Core Laboratory at Texas A&M University, as well as equipment and supplies needed for the program. The MN Zoo's bison conservation program is coordinated by a dedicated biologist with additional support provided by other members of the Conservation team as well as the Animal Health and Animal Care staff.

3. Lessons Learned

Wildlife conservation can be accomplished through many tactics, including both in situ and ex situ programs. Different institutions are best positioned to contribute to conservation in specific ways depending on their expertise and available resources. Zoos and aquariums are unique players in the conservation field today, due in part to their specialized facilities, staff expertise in ex situ conservation programs and animal care and health, access to large audiences, and diverse animal collections. As zoos and aquariums continue moving towards conservation-minded missions and initiatives that leverage their expertise and partnerships, there may be increased opportunities to contribute directly to local wildlife conservation [53]. By examining the history and current status of the MN Zoo's four main local conservation programs, and considering our unique status as a state agency, we summarize a few key lessons learned from our own experiences.

Our status as a state agency necessitates our focus on local wildlife.

As a state agency, the MN Zoo is committed to serving Minnesotans by bringing our mission to life through the conservation of wildlife within the state and by connecting people to nature. Much of our field conservation work is directly supported by state funds, and our current portfolio of projects remains largely focused within Minnesota. We are orchestrating five significant field conservation projects and four of these (pollinators, turtles, freshwater mussels, and bison) are almost entirely located in Minnesota; three of these programs are funded in large part by the state's ENRTF. We also currently coordinate a black rhino conservation program in Namibia, but this initiative is funded exclusively by the Minnesota Zoo Foundation, the affiliated 501c3 that helps provide support to the MN Zoo, since state funds cannot be used to support an international project.

Additionally, as a government agency, we receive a bi-annual appropriation from the state that helps supports our programs. Of course, species-focused conservation work often entails needing to work across states and even countries, which has compelled us to look for additional financial support. Soon after the Poweshiek skipperling work at the MN Zoo began, the remaining populations within the state disappeared. We thus have secured funding from organizations outside the state to continue working with the remaining population in Michigan.

Partnerships are essential for fledgling, zoo-based conservation initiatives and necessary to advance institutional field conservation objectives.

One of the most influential factors for the MN Zoo's conservation work has been our standing and newly formed partnerships. Not only have our partnerships been hugely influential as our conservation programs have grown, but many of our current projects likely would not have started without them. The MN Zoo's pollinator, turtle, native mussel, and bison initiatives all came to fruition or were shaped dramatically by standing relationships, most notably with the DNR. The identified conservation needs and how best we could leverage our assets as a zoo-based conservation organization informed our initial roles. When initially structuring our conservation programs, we found it most logical and productive to support existing programs through our partnerships rather than create new programs entirely with our own resources. Perhaps surprisingly, the standing collection of animals at the MN Zoo at the time these programs were created had little to no bearing on how these conservation initiatives took form.

Today, our Minnesota-based conservation programs continue to benefit greatly from a robust portfolio of partnering institutions. While we are fortunate to have access to the natural landscapes of the MN Zoo, our jurisdiction ends there. In order to conduct off-site surveys and species reintroductions, we need to collaborate with a variety of institutions. This is evident in all our field projects. The reintroductions and surveys for wood turtles, prairie butterflies, freshwater mussels, and bison all rely on the cooperation of other organizations and individuals. To structure and conduct effective conservation programs, both in situ and ex situ efforts should be considered, and for our programs, this has only been made possible through a collaborative approach.

As an agency of the State of Minnesota, we are well-positioned to partner with other government agencies to benefit conservation.

Many of our partners also operate under the state's jurisdiction. The Minnesota Department of Natural Resources, another state agency, is a key partner across all our field initiatives. We also maintain close relationships with the Departments of Transportation and Corrections, as well as with the University of Minnesota, which is the largest public university system in the state. In several ways, our status as a state agency has facilitated and simplified our ability to work with other large state institutions. As we operate under many of the same guidelines and procedures, formalizing partnerships and agreements is often simpler and faster than if we were entering into similar partnerships with an organization outside the state system. It has also promoted our ability to work with other government entities including the U.S. Fish and Wildlife Service and county governments. State agency status has even created mandates for conservation partnerships with other agencies. For instance, the MN Zoo has been directed under Minnesota Governor's Executive Orders 16-07 [54] and 19-28 [55] to be a member of the Interagency Pollinator Protection Team, a group of ten state agencies convened to advance pollinator conservation and engagement in Minnesota. This membership has necessitated staff time but also has fostered relationships and decreased silos with other agencies.

Programmatic flexibility and opportunism have been key in maximizing growth potential.

In examining our portfolio of current projects, it is evident how significantly our work has grown and pivoted at times throughout the past decade. The Pollinator Conservation Initiative was created to help assess the current status of two potentially declining butterfly species, and we now have pioneered and implemented a robust rearing and breeding program for both the Dakota skipper and Poweshiek skipperling, working with partners across the continent to save them from extinction. Our freshwater turtle conservation work continues to grow as we now have a well-established head-starting program for wood turtles and are expanding efforts to encompass the threatened Blanding's turtle as well. The native mussel program is rearing more individuals and new species, including walleye, in facilities at the MN Zoo and in partnership with the DNR, and now has a new focus on the invasive zebra mussel. Lastly, the Minnesota Bison Conservation Herd continues to seek out and create new partnerships as it works to meet its goal of a statewide collective herd of 500 animals.

Over the years, we have found it necessary to consistently reevaluate the conservation landscape, maintain communication with our partners, and reassess how the MN Zoo can best support those efforts. Shifting conservation priorities and the changing capacities of our partners have frequently warranted a directional change within our own programs. The growth and expansion of these programs is exciting, and the ability to shift focus as the ever-changing needs develop is imperative to their success.

The Minnesota Zoo's ability to procure external grants to support its conservation initiatives has resulted in programmatic siloing.

As is the case with many other conservation programs, the various funding streams that we have received over the years have significantly impacted our conservation efforts. Currently, much of our work is supported through Minnesota's Environment and Natural Resources Trust Fund (ENRTF). The money in this fund is generated by the State Lottery and administered each year by the Legislative-Citizen Commission on Minnesota's Resources Since 1991, the ENRTF has funded approximately $700 million to thousands of projects across the state [28].

We have found that having projects sustained by external grants can create a silo effect. Since staff time is largely supported by grants to work on specific projects and must be reported on periodically, there is typically little flexibility for departmental staff to work and help with other field projects which would be especially desirable during the busy field season. It is also challenging to build institutional knowledge and ensure programmatic continuity when staff are mostly confined to working within their programs. The specifics of the grant and the awarding institution can additionally influence how a conservation project functions. A change in funding may result in a necessary pivot in one way or another. As our conservation programs continue to grow, maintaining programmatic flexibility will be essential to their success. Transitioning toward a model that is less reliant on external grant funding will improve efficiency and programmatic sustainability and effectiveness. Specifically, we hope to support the salaries of full-time Conservation staff via the MN Zoo's general operating budget and use external grants to fund project-specific costs, such as equipment and supplies, travel, and temporary staff that may be necessary for implementation.

However, grant funding can buffer against the impacts of economic uncertainty and severe downturns.

Relying on grants as a main source of financial support typically lends some instability from year to year as exact future funds are unknown. During the initial phase of the COVID-19 pandemic, however, our reliance on grant funding offered our department rare and unexpected stability. As with many institutions across the globe, the MN Zoo was forced to close its doors for months in response to the pandemic which necessitated lay-offs for a significant portion of the workforce and dramatic spending cuts. Because our funding came from external sources outside of the MN Zoo's general operating budget, our local conservation programs were spared from staff layoffs and many other budgetary cuts.

4. Conclusions

By reviewing and analyzing the history and current status of our conservation initiatives, we can piece together how the MN Zoo has leveraged its unique strengths and assets to impact local wildlife conservation. Similar to other zoos, the MN Zoo remains well-positioned to aid in ex situ efforts, especially in the formation and execution of head-starting and similar programs, largely due to our facilities and staff expertise across the organization.

In addition, our designation as a state agency, rare among zoos and aquariums, has shifted our focus towards local efforts as a result of the financial support received from State of Minnesota sources and our ability to partner with other government agencies. This local focus is not necessarily typical for AZA institutions. Whereas about 90% of AZA member institutions are located in the US [9], only 45% of field conservation projects took place within the country in 2021 [56]. By contrast, 80% of the MN Zoo's major field research projects take place entirely within the US (and primarily within Minnesota). As our focus has localized and changed over the years, we have learned that maintaining a degree of programmatic flexibility is key to the successful growth and impact of our conservation initiatives.

We suggest that other institutions consider focusing a larger portion of their conservation portfolios on local initiatives; local wildlife also may require conservation attention and may represent a more efficient investment of limited resources. Moreover, our experiences suggest that local conservation impact can resonate with zoo visitorship, even if the target species are not considered charismatic megafauna (in our case, butterflies, mussels, and turtles). Although our role as a state agency has helped facilitate the partnerships necessary to engage in local conservation, zoos and aquariums without that standing can familiarize themselves and engage with their unique local conservation partners to ensure an impactful and collaborative approach.

Constructing and understanding our own narrative is valuable in considering what the MN Zoo's initiatives may look like in the future so that we can build our efforts to have a significant and lasting conservation impact. As we have learned through the development and growth of our own local conservation programs, zoos and aquariums can play a critical role in wildlife conservation. Every institution, however, has its own unique background, and individual strengths and narratives are likely to vary considerably among organizations based on that context. We encourage other zoos and aquariums to undertake a similar analysis of their own programs when setting future conservation goals and priorities.

The role of zoos and aquariums has changed significantly over time, from entertainment centers focused on providing the public with a view of often rare and exotic animals to collaborative institutions with a mission to save wildlife and champion animal welfare [57]. In the fight to save wildlife, the global community must utilize every tool at its disposal, leveraging the distinct assets that each organization brings to the table, including zoos and aquariums.

Author Contributions: Conceptualization, S.S. and M.M.; writing—original draft preparation, M.M. and S.S.; writing—reviewing and editing, M.M., T.M., B.M., C.N., E.R. and S.S. All authors have read and agreed to the published version of the manuscript.

Funding: Funding and in-kind support for the conservation work of the Minnesota Zoo reviewed here is provided by numerous organizations, including Minnesota's Environment and Natural Resources Trust Fund; Association of Zoos and Aquariums; Disney Conservation Fund; Great Lakes Restoration Initiative; Minnesota's Clean Water, Land, and Legacy Amendment; Minnesota Department of Transportation; Minnesota Zoo Foundation; Minnesota Zoological Gardens; National Geographic; Morrie's Automotive Group; North Dakota Natural Resources Trust; Shakopee Mdewakanton Sioux Community; and the University of Minnesota's Minnesota Aquatic Invasive Species Research Center.

Institutional Review Board Statement: Not applicable.

Informed Consent Statement: Not applicable.

Data Availability Statement: Not applicable.

Acknowledgments: We are grateful to the many individuals who have contributed to the Minnesota Zoo's conservation initiatives throughout the years, especially departmental interns and student workers, our colleagues at the Minnesota Zoo and Minnesota Zoo Foundation, and Minnesota Zoo volunteers. We also thank our numerous partners across Minnesota and beyond for their collaboration and tireless dedication to improving the conservation of local wildlife for future generations.

Conflicts of Interest: The authors declare no conflict of interest.

References

1. Redford, K.H.; Coppolillo, P.; Sanderson, E.W.; da Fonseca, G.A.B.; Dinerstein, E.; Groves, C.; Mace, G.; Maginnis, S.; Mittermeier, R.A.; Noss, R.; et al. Mapping the conservation landscape. *Conserv. Biol.* **2003**, *17*, 116–131. [CrossRef]
2. Braverman, I. Conservation without nature: The trouble with in situ versus ex situ conservation. *Geoforum* **2014**, *51*, 47–57. [CrossRef]
3. Mace, G.; Balmford, A.; Boitani, L.; Cowlishaw, G.; Dobson, A.P.; Faith, D.P.; Gaston, K.J.; Humphries, C.J.; Vane-Wright, R.I.; Williams, P.H. It's time to work together and stop duplicating conservation efforts. *Nature* **2000**, *405*, 393. [CrossRef] [PubMed]
4. Conservation Planning Specialist Group [CPSG]. 2022. Available online: http://cpsg.org/our-approach/one-plan-approach-conservation (accessed on 18 October 2022).
5. Association of Zoos and Aquariums [AZA]. Interesting Zoo and Aquarium Statistics. 2021. Available online: https://www.aza.org/connect-stories/stories/interesting-zoo-aquarium-statistics?locale=en (accessed on 19 October 2022).
6. Bouman, I.; Wit, P. *The Tale of the Przewalski's Horse: Coming Home to Mongolia*; KNNV Publishers: Utrecht, The Netherlands, 2006.
7. Xia, C.; Cao, J.; Zhang, H.; Gao, X.; Yang, W.; Blank, D. Reintroduction of Przewalski's horse (Equus ferus przewalskii) in Xinjiang, China: The status and experience. *Biol. Conserv.* **2014**, *177*, 142–147. [CrossRef]
8. Association of Zoos and Aquariums [AZA]. About Us. 2022. Available online: https://www.aza.org/about-us?locale=en (accessed on 20 October 2022).
9. Association of Zoos and Aquariums [AZA]. Zoo and Aquarium Statistics. 2022. Available online: https://www.aza.org/zoo-and-aquarium-statistics (accessed on 20 October 2022).
10. Samson, F.; Knopf, F. Prairie conservation in North America. *BioScience* **1994**, *44*, 418–421. [CrossRef]
11. White, R.; Murray, S.; Rohwede, M. *Pilot Analysis of Global Ecosystems: Grassland Ecosystems*; World Resources Institute: Washington, DC, USA, 2000.
12. Brennan, L.A.; Kuvlesky, W.P. North American grassland birds: An unfolding conservation crisis? *J. Wildl. Manag.* **2005**, *69*, 1–13. [CrossRef]
13. Leach, M.K.; Givnish, T.J. Ecological determinants of species loss in remnant prairies. *Science* **1996**, *273*, 1555–1558. [CrossRef]
14. Miles, E.K.; Knops, J.M.H. Grassland compositional change in relation to the identity of the dominant matrix-forming species. *Plant Ecol. Divers.* **2009**, *2*, 265–275. [CrossRef]
15. Cochrane, J.F.; Delphey, P. *Status Assessment and Conservation Guidelines: Dakota Skipper (Hesperia Dacotae)*; U.S. Fish and Wildlife Service: Washington, DC, USA, 2002.

16. McCabe, T.L. The Dakota skipper, Hesperia dacota (Skinner): Range and biology with special reference to North Dakota. *J. Lepid. Soc.* **1981**, *35*, 179–193.
17. Schlicht, D.; Swengel, A.; Swengel, S. Meta-analysis of survey data to assess trends of prairie butterflies in Minnesota, USA during 1979–2005. *J. Insect Conserv.* **2009**, *13*, 429–447. [CrossRef]
18. Swengel, A.B.; Swengel, S.R. Grass-skipper (Hesperiinae) trends in Midwestern USA grasslands during 1988–2013. *J. Insect Conserv.* **2015**, *19*, 279–292. [CrossRef]
19. U.S. Fish and Wildlife Service [USFWS]. *Endangered and Threatened Wildlife and Plants; Threatened Species Status for Dakota Skipper and Endangered Species Status for Poweshiek Skipperling;* 50 CFR Part 17; Final Rule; U.S. Fish and Wildlife Service: Washington, DC, USA, 2014; pp. 63672–63748.
20. Royer, E. *Hesperia dacotae. The IUCN Red List of Threatened Species: E.T9968A122963341;* IUCN Global Species Programme Red List Unit: Cambridge, UK, 2019. [CrossRef]
21. Royer, E. *Oarisma poweshiek. The IUCN Red List of Threatened Species: E.T122914337A166163683;* IUCN Global Species Programme Red List Unit: Cambridge, UK, 2020. [CrossRef]
22. IUCN/SSC. *Guidelines on the Use of Ex Situ Management for Species Conservation;* Version 2.0; IUCN Species Survival Commission: Gland, Switzerland, 2014.
23. Delphey, P.; Runquist, E.; Harris, T.; Nordmeyer, C.; Smith, T.; Traylor-Holzer, K.; Miller, P.S. *Poweshiek Skipperling and Dakota Skipper: Ex Situ Feasibility Assessment and Planning Workshop;* IUCN/SSC Conservation Breeding Specialist Group: Apple Valley, MN, USA, 2016.
24. Crone, E.E.; Pickering, D.; Schultz, C.B. Can captive rearing promote recovery of endangered butterflies? An assessment in the face of uncertainty. *Biol. Conserv.* **2007**, *139*, 103–112. [CrossRef]
25. Thomas, J.A.; Simcox, D.J.; Hovestadt, R.T. Evidence based conservation of butterflies. *J. Insect Conserv.* **2011**, *15*, 241–258. [CrossRef]
26. IUCN/SSC. *Guidelines for Reintroductions and Other Conservation Translocations;* Version 1.0; IUCN Species Survival Commission: Gland, Switzerland, 2013.
27. Nordmeyer, C.S.; Runquist, E.; Stapleton, S. Invasive grass negatively affects growth and survival of an imperiled butterfly. *Endanger. Species Res.* **2021**, *45*, 301–314. [CrossRef]
28. Minnesota Legislative Coordinating Commission. Environment & Natural Resources Trust Fund. 2022. Available online: https://www.legacy.mn.gov/environment-natural-resources-trust-fund (accessed on 24 October 2022).
29. Minnesota Department of Natural Resources [DNR]. Minnesota Wood Turtle Conservation Plan. 2020. Available online: https://files.dnr.state.mn.us/eco/nongame/projects/mn-wood-turtle-conservation-plan (accessed on 13 October 2022).
30. Lovich, J.E.; Ennen, J.R.; Agha, M.; Gibbons, J.W. Where have all the turtles gone, and why does it matter? *BioScience* **2018**, *68*, 771–781. [CrossRef]
31. Howell, H.J.; Legere, R.H., Jr.; Holland, D.S.; Seigel, R.A. Long-term turtle declines: Protected Is a verb, not an outcome. *Copeia* **2019**, *107*, 493–501. [CrossRef]
32. Walde, A.D.; Bider, J.R.; Masse, D.; Saumure, R.A.; Titman, R.D. Nesting ecology and hatching success of the wood turtle, *Glyptemys inscultpa*, in Quebec. *Herpetol. Conserv. Biol.* **2007**, *2*, 49–60.
33. Wallace, S.D.; Forbes, G.J.; Nocera, J.J. Experimental assessment of the impact of agricultural machinery on wood turtles (*Glyptemys insculpta*). *Chelonian Conserv. Biol.* **2020**, *19*, 78–84. [CrossRef]
34. U.S. Fish and Wildlife Service [USWFWS]. Wood Turtle. 2022. Available online: https://www.fws.gov/species/wood-turtle-glyptemys-insculpta (accessed on 12 October 2022).
35. U.S. Fish and Wildlife Service [UWFWS]. Blanding's Turtle. 2022. Available online: https://www.fws.gov/species/blandings-turtle-emydoidea-blandingii (accessed on 12 October 2022).
36. Minnesota Department of Natural Resources [DNR]. Blanding's Turtle Survey and Conservation. 2022. Available online: https://www.dnr.state.mn.us/eco/nongame/projects/blandings_survey_conservation.html (accessed on 14 October 2022).
37. Markel, T.; Stapleton, S. *Reduce Vehicle-Animal Collisions with Installation of Small Animal Exclusion Fencing;* Research Report; Minnesota Department of Transportation: Saint Paul, MN, USA, 2022.
38. van Dijk, P.P.; Rhodin, A.G.J. *Emydoidea blandingii. (Errata Version Published in 2016). The IUCN Red List of Threatened Species: E.T7709A97411815;* IUCN Global Species Programme Red List Unit: Cambridge, UK, 2011.
39. Pappas, M.J.; Brecke, B.J.; Congdon, J.D. The Blanding's turtle of Weaver Dunes, Minnesota. *Chelonian Conserv. Biol.* **2000**, *3*, 557–568.
40. Helfrich, L.A.; Neves, R.J.; Chapman, H. *Sustaining America's Aquatic Biodiversity: Freshwater Mussel Biodiversity and Conservation;* Virgina Cooperative Extension; Virgina Tech: Blacksburg, VA, USA, 2019; pp. 420–523.
41. Neves, R.J. Propagation of endangered freshwater mussels in North America. *J. Conchol.* **2004**, *3*, 69–80.
42. Strayer, D.L.; Downing, J.A.; Haag, W.R.; King, T.L.; Layzer, J.B.; Newton, T.J.; Nichols, J.S. Changing perspectives on pearly mussels, North America's most imperiled animals. *BioScience* **2004**, *54*, 429–439. [CrossRef]
43. Kelly, M. *Species Profile: Freshwater Mussels, Sentinals of the Water;* Minnesota Department of Natural Resources: St. Paul, MN, USA, 2012.
44. Minnesota Department of Natural Resources [DNR]. Mussels of Minnesota. 2022. Available online: https://www.dnr.state.mn.us/mussels/index.html (accessed on 2 November 2022).

45. Patterson, M.A.; Mair, R.A.; Eckert, N.L.; Gatenby, C.M.; Brady, T.; Jones, J.W.; Simmons, B.R.; Devers, J.L. *Freshwater Mussel Propagation for Restoration;* Cambridge University Press: Cambridge, UK, 2018.
46. Strayer, D.L. Twenty years of zebra mussels: Lessons from the mollusk that made headlines. *Front. Ecol. Environ.* **2008**, *7*, 135–141. [CrossRef]
47. Boyd, D.P.; Gates, C.C. A brief review of the status of plains bison in North America. *J. West* **2006**, *45*, 15–21.
48. Freese, C.H.; Aune, K.E.; Boyd, D.P.; Derr, J.N.; Forrest, S.C.; Gates, C.C.; Gogan, P.J.P.; Grassel, S.M.; Halbert, N.D.; Kunkel, K.; et al. Second chance for the plains bison. *Biol. Conserv.* **2007**, *136*, 175–184. [CrossRef]
49. Steenweg, R.; Hebblewhite, M.; Gummer, D.; Low, B.; Hunt, B. Assessing potential habitat and carrying capacity for reintroduction of plains bison (Bison bison bison) in Banff National Park. *PLoS ONE* **2016**, *11*, e0150065. [CrossRef]
50. Stroupe, S.; Forgacs, D.; Harris, A.; Derr, J.N.; Davis, B.W. Genomic evaluation of hybridization in historic and modern North American bison (Bison bison). *Sci. Rep.* **2022**, *12*, 6397. [CrossRef]
51. Hedrick, P.W. Conservation genetics and North American bison (Bison bison). *J. Hered.* **2009**, *100*, 411–420. [CrossRef]
52. Minnesota Department of Natural Resources [DNR]- Division of Parks and Trails. Strategic Plan for Bison Management. 2016. Available online: https://files.dnr.state.mn.us/input/mgmtplans/parks/bison_strategic%20plan_121516_final.pdf (accessed on 3 October 2022).
53. Keulartz, J. Captivity for conservation? Zoos at a crossroads. *J. Agric. Environ. Ethics* **2015**, *28*, 335–351. [CrossRef]
54. State of Minnesota—Executive Department. Executive Order 16-07: Directing Steps to Reverse Pollinator Decline and Restore Pollinator Health in Minnesota. Minnesota Legislative Reference Library. 2016. Available online: https://www.leg.mn.gov/archive/execorders/16-07.pdf (accessed on 7 November 2022).
55. State of Minnesota—Executive Department. Executive Order 19-28: Restoring Healthy, Diverse Pollinator Populations That Sustain and Enhance Minnesota's Environemnt, Economy, and Way of Life. Minnesota Legislative Reference Library. 2019. Available online: https://mn.gov/governor/assets/2019_04_05_EO_19-28_tcm1055-379048.pdf (accessed on 7 November 2022).
56. Association of Zoos and Aquariums [AZA]. Annual Report on Conservation and Science: Highlights. 2021. Available online: https://assets.speakcdn.com/assets/2332/aza_arcshighlights_2021_final_web.pdf (accessed on 22 February 2023).
57. Tribe, A.; Booth, R. Assessing the role of zoos in wildlife conservation. *Hum. Dimens. Wildl.* **2006**, *8*, 65–74. [CrossRef]

Journal of
*Zoological and
Botanical Gardens*

Review

Conservation through Collaboration: Regional Conservation Programs of the North Carolina Zoo

Elizabeth A. Roznik *, Halley Buckanoff, Robert W. Langston, Christopher J. Shupp and Dustin Smith

North Carolina Zoo, Asheboro, NC 27205, USA
* Correspondence: betsy.roznik@nczoo.org

Abstract: In response to rapid biodiversity losses in recent decades, zoos have become more engaged in conservation issues. Solutions to conservation challenges are complex and require collaborative efforts across organizations. Zoos can be effective partners that can contribute diverse expertise and resources to protect wildlife and their habitats. While zoos often partner with international organizations to facilitate field-based conservation projects on the exotic animals they exhibit, some of the most meaningful conservation and education initiatives are conducted locally in partnership with local organizations. A core part of the mission of the North Carolina Zoo (Asheboro, NC, USA) is the conservation of wildlife and their natural habitats, both regionally and internationally. The goal of this article is to review the North Carolina Zoo's regional conservation programs and the importance of partnerships with other local organizations in accomplishing shared goals. North Carolina Zoo plays an important role in regional conservation by protecting and managing natural lands, protecting declining amphibians through headstarting and habitat management, rehabilitating native wildlife, and working on local outreach and sustainability projects to reduce impacts on natural resources and inspire others to get involved in conservation. These programs were developed through partnerships with local and state government agencies, academic institutions, non-profit organizations, other zoos and aquariums, schools, libraries, and businesses. These collaborations have been instrumental in developing and implementing successful projects by pooling limited resources and sharing crucial expertise. They demonstrate how zoos are evolving to become leaders and partners in conservation, research, and education to protect local species and natural resources.

Keywords: collaboration; conservation; headstarting; land management; outreach; rehabilitation; sustainability; zoos

Citation: Roznik, E.A.; Buckanoff, H.; Langston, R.W.; Shupp, C.J.; Smith, D. Conservation through Collaboration: Regional Conservation Programs of the North Carolina Zoo. *J. Zool. Bot. Gard.* **2023**, *4*, 292–311. https://doi.org/10.3390/jzbg4020025

Academic Editors: Ursula Bechert and Debra C. Colodner

Received: 15 October 2022
Revised: 15 February 2023
Accepted: 10 March 2023
Published: 24 March 2023

1. Introduction

Global biodiversity is under severe threat from human impacts, including habitat loss, pollution, overexploitation, emerging diseases, invasive species, and climate change. In the past 500 years, humans have triggered a wave of extinction and decline that may be comparable in both rate and magnitude to the five previous mass extinctions in Earth's history [1,2]. The rate of this sixth mass extinction has been estimated as 1000 times the background rate of extinction [3], and over one-third of all known vertebrates are decreasing in population size and geographic range [4]. These losses will have negative cascading consequences on ecosystem functioning and ecological services.

Slowing down biodiversity loss is one of the greatest challenges facing humans today. While many groups are working to address threats to biodiversity, such as government agencies, academic institutions, non-profit organizations, and private landowners, collaboration between organizations is often important for achieving shared goals. Collaborative conservation includes efforts to preserve, protect, and/or sustainably manage natural resources by two or more partners working together, often to set goals, make decisions, and implement actions [5]. Collaborative efforts are increasingly recognized as important for addressing landscape-scale issues, and they are valuable for sharing resources, expertise, and diverse perspectives that drive innovation [6].

During the last 50 years, zoos have evolved from serving as entertainment venues to becoming centers for conservation, research, and education [7–9]. A key driver of this change was the development of captive-breeding programs first implemented to sustainably manage populations housed in zoos [9], and later applied to captive-breeding and release programs for threatened species. Thus, in response to growing biodiversity losses, zoos have become more engaged in conservation issues. Annually, zoos and aquariums accredited by the Association of Zoos and Aquariums (AZA) spend over USD 217 million on field conservation with more than 960 partners [10], and host over 183 million visitors [11]. This demonstrates the opportunities for these institutions to play an important role in environmental education and conservation of wildlife and their natural habitats. Zoos have become valuable partners in the recovery of threatened species by contributing to the care, research, and genetic management of zoo-managed populations, facilitating captive-breeding and reintroduction programs, and funding and participating in field-based conservation efforts [7,8,12]. Thus, conservation activities can occur in situ, ex situ, or both, and innovative solutions that bridge or blur these concepts are becoming increasingly important in species conservation [13]. While zoos often partner with international organizations to fund and assist with field-based conservation and research projects on the exotic animals they exhibit, some of the most meaningful conservation and education initiatives are conducted locally in partnership with local organizations.

A core part of the mission of the North Carolina Zoo (NCZ) is to conserve wildlife and their natural habitats. The scope of NCZ's conservation activities is geographically and taxonomically broad. Internationally, some of the current projects focus on protecting vultures and gorillas in Africa, translocating rare birds on Pacific islands, and developing software that empowers rangers worldwide in the fight against the illegal wildlife trade. Closer to home, NCZ plays an important role in regional conservation by protecting and managing natural lands, protecting threatened amphibians through headstarting and habitat management, rehabilitating native wildlife, and working on local outreach and sustainability projects to reduce NCZ's impact on natural resources and inspire guests and community members to get involved in conservation. These regional conservation programs were developed through partnerships with local and state agencies and other organizations. These collaborations have been instrumental in developing and implementing successful projects by pooling limited resources and sharing crucial expertise.

The purpose of this article is to review NCZ's regional conservation programs and the importance of partnerships with other local conservation organizations in accomplishing shared goals that could not be achieved by working alone. Although NCZ has made many other significant contributions to conservation and research, including international and zoo-based initiatives, we focus this article on conservation of native species and natural resources in North Carolina. We also use our experiences to provide recommendations for other zoos to begin or become involved with local conservation initiatives.

2. North Carolina Zoo

North Carolina Zoo was established in Asheboro, North Carolina, USA, in 1974. As a state-supported zoo, it is an agency in the Department of Natural and Cultural Resources. While the state provides an operating budget, the North Carolina Zoological Society is a non-profit partner that raises funds for conservation programs, capital campaigns, and other initiatives through memberships, donations, special events, gift shop sales, and other fundraising activities. NCZ welcomes up to one million guests each year. NCZ was established as the world's largest natural habitat zoo to provide large spaces for animals in natural settings. With this goal, the developed portion of NCZ containing exhibits and support structures (200 ha) was placed on a large tract of land surrounded by over 400 ha of natural land on Purgatory Mountain (Figure 1). Early explorers suggested that the region was occupied by Waxhaw, Sugeree, Saponi, Saura, Wateree, and/or Catawba native peoples that were displaced by the influx of Europeans by 1750 [14]. More recently, the property was used for mining activities, and according to local lore, Purgatory Mountain was used as a liquor distillery and named for the numerous fires dotting the hillside during bootlegging operations [14]. NCZ houses more than 1700 animals representing North America and Africa, with construction of a new Asia exhibit currently underway. While NCZ has always been committed to the conservation of wildlife and their natural habitats, this focus has grown steadily over the years through the expansion of conservation and education programs supported by increased funding and staff positions. Funding increased from USD 248,000 in 2007 to USD 1.1 million in 2021. The Conservation, Education, and Science section of NCZ currently comprises 18 permanent staff, including five staff dedicated primarily to conservation and research. Animal management staff also strongly contribute to regional and international conservation programs, and additional support is provided by many seasonal staff, interns, AmeriCorps members, and volunteers.

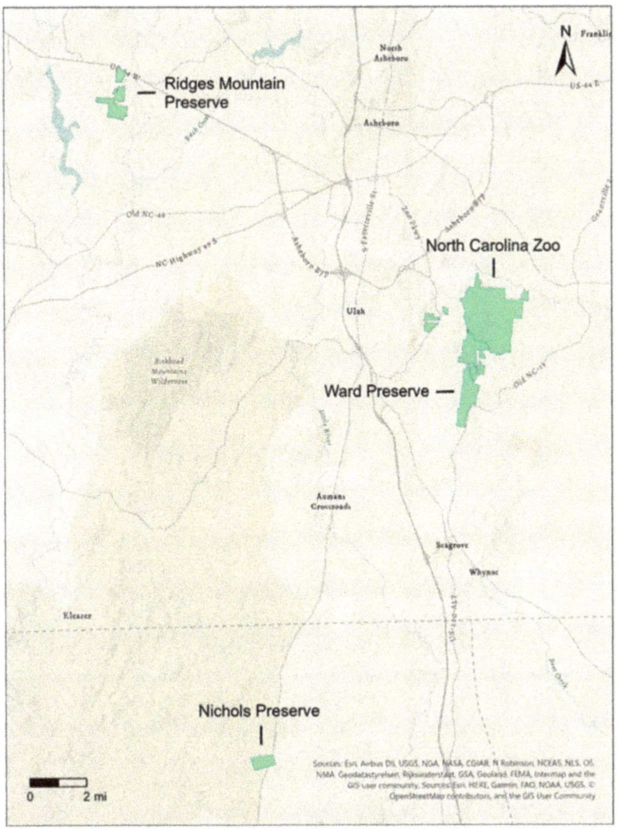

Figure 1. North Carolina Zoo plays an important role in land conservation by protecting and managing more than 800 ha of undeveloped land in central North Carolina. This includes land on and surrounding NCZ's main property and three off-site nature preserves: Nichols Preserve, Ridges Mountain Preserve, and Ward Preserve. The shaded region to the west of NCZ is the Uwharrie National Forest (containing the Birkhead Mountains Wilderness), which is the largest public landholder in the region.

3. Land Conservation and Management

3.1. Overview

Biodiversity is under threat worldwide and protected natural areas are widely recognized for their value in biodiversity conservation [15,16]. Because the greatest threats to biodiversity are landscape modification and fragmentation, the most effective protected areas preserve large contiguous tracts of high-quality natural communities [17,18]. While many protected areas are established to protect rare species or communities, common species also benefit from land protection and are often present in higher densities and demonstrate greater temporal stability [19]. Common species provide vital ecosystem processes and are useful indicators of ecosystem function and health [20,21]. In addition to supporting biodiversity, many natural areas also provide benefits for people, especially areas that provide recreational trails. Hiking and other outdoor activities provide opportunities for people to improve their mental and physical health and develop a deeper appreciation and sense of stewardship for protected natural areas [22–24].

North Carolina Zoo plays an important role in land conservation by protecting and managing more than 800 ha of undeveloped land in central North Carolina (Figures 1 and 2). NCZ is located in a rural area approximately 11 km from the city of Asheboro, which has a population of 25,000 people. The majority of public land in the region is part of the Uwharrie National Forest (20,860 ha), with most remaining land under private ownership. There is ongoing loss, degradation, and fragmentation of natural habitat in the region due to agriculture and urban development. The natural areas protected by the NCZ encompass a diversity of ecosystems and protect unique biological, geological, and cultural features. They also provide unfragmented spaces and movement corridors that allow wildlife to safely navigate the landscape. NCZ was established on a large tract of land surrounded by over 400 ha of natural land on and around Purgatory Mountain (285 m) at the northern edge of the Uwharrie range. This land supports a diversity of natural communities, including some that are rare in the region, such as upland pools that provide valuable breeding sites for amphibians, such as the four-toed salamander (*Hemidactylium scutatum*), which is species of special concern in North Carolina. Because of its size and habitat heterogeneity, this natural area provides important wildlife habitat, especially for forest interior species that require large areas of unfragmented forest.

NCZ expanded its land conservation footprint from its original property by acquiring additional land in the vicinity to create three off-site nature preserves. Ridges Mountain Nature Preserve (Ridges Mountain Preserve, 75 ha, established in 2000) protects Ridges Mountain (256 m) in the northernmost range of the ancient Uwharrie Mountains. The geology of Ridges Mountain is striking because of the enormous boulders, up to 15 m in height, that are found along the ridge line. In addition to preserving unique geology, Ridges Mountain Preserve protects high-quality mature forests and wetlands, and supports an assemblage of uncommon plant species, such as fragrant sumac (*Rhus aromatica*), due to basic soil that differs from the more typical acidic soils in the region. On a landscape scale, Ridges Mountain is also significant due to its connectivity to other large, forested areas along an upland to bottomland corridor. The Selma Cornelison Ward Nature Preserve (Ward Preserve, 131 ha, established in 2010) provides mature hardwoods along Bachelor Creek. This preserve protects a forested corridor between other forested areas in the region, including NCZ's main property, and it buffers the headwaters of the creek, supporting water quality and flow downstream. The Margaret J. Nichols Longleaf Pine Forest Preserve (Nichols Preserve, 47 ha, established in 2011) protects the largest remaining known stand of old-growth longleaf pine (*Pinus palustris*) in the Piedmont of North Carolina, containing trees that are more than 200 years old. The longleaf pine ecosystem, which is one of the most biodiverse ecosystems on the planet and contains many endemic species, has declined by more than 97% across its range in the southeastern USA [25].

Figure 2. North Carolina Zoo plays an important role in land conservation by (**A**) protecting ecologically significant lands, including this upland pool that provides important breeding habitat for amphibians, and by (**B**) managing these lands to maintain or improve their quality, such as by applying prescribed fire to this longleaf pine forest in partnership with the North Carolina Forest Service.

3.2. Land Protection

NCZ focuses on protecting larger tracts of unfragmented land that support high-quality natural communities (Figure 2A). Valuable partnerships with other state agencies and local non-profit organizations have assisted with identifying and facilitating the protection of ecologically significant land. The North Carolina Natural Heritage Program (NCNHP) collects and shares information about rare species and natural communities that is needed to evaluate the ecological significance of natural areas and potential ecological impacts of conservation and development projects. County-level natural area inventories conducted by the NCNHP were key in identifying the high-quality natural communities present on NCZ's main property and those protected on the three off-site nature preserves.

Once ecologically valuable lands were surveyed and identified, partnerships with the Three Rivers Land Trust and Piedmont Land Conservancy were key in locating land in need of protection and facilitating their acquisition. NCZ and Three Rivers Land Trust worked together to protect the land that became the Nichols Preserve. NCZ acquired half of the property by obtaining grant funds through the North Carolina Natural Heritage Trust Fund, which was established to preserve natural areas and rare species across the state.

The Three Rivers Land Trust purchased the other half of the property using interest-free loan funding through the Norcross Wildlife Foundation, and then worked with NCZ to obtain additional funds from the North Carolina Natural Heritage Trust Fund to sell the remainder of the property to NCZ. Piedmont Land Conservancy provided support and funding for acquiring the Ridges Mountain Preserve, along with funding provided by the landowner, Mary and Elliot Wood Foundation, Cannon Foundation, Kathleen Price Bryan Family Fund, Hans Klaussner Foundation, and the North Carolina Natural Heritage Trust Fund. Piedmont Land Conservancy also provided legal assistance for acquiring the Ward Preserve, which was purchased using grant funding from the North Carolina Natural Heritage Trust Fund.

3.3. *Land Management*

Effective land management is essential for successful habitat and species conservation. NCZ manages natural areas to maintain high quality or increase quality where possible. The partners involved with acquiring NCZ's preserves continue to be involved with management planning, while NCZ serves as the steward and implements the land management plans. Detailed management plans have been written for all natural areas that include the history, significance, goals, and current and planned management practices. Several biological consultants have been instrumental in assisting with land management planning and support, particularly on the Nichols Preserve. An independent consultant, Terry Sharpe (Certified Wildlife Biologist and NC Registered Forester), created a detailed management plan for restoring the degraded old-growth longleaf pine forest on the site. Eli Beverly and Associates, LLC, has provided significant advice and expertise in managing invasive plant species.

While NCZ staff conduct routine maintenance in natural areas, such as mowing around parking areas and clearing hiking trails, more intensive management is conducted through partnerships. One example is prescribed fire, which is a management tool used to maintain and improve fire-adapted communities, such as the longleaf pine forest on the Nichols Preserve. NCZ partners with the North Carolina Forest Service to conduct prescribed burns on this site and other NCZ properties (Figure 2B). Trained burn crews with the North Carolina Forest Service carefully plan and conduct prescribed burns by considering safety, weather, and management goals. NCZ also works with volunteers to accomplish other management activities on natural areas, such as invasive plant control and litter cleanups.

NCZ staff conduct regular surveys and research studies to assist with conservation and management planning. This includes surveying plant, bird, mammal, amphibian, reptile, and invertebrate communities to inventory rare species and monitor populations over time. For example, one study documented the species diversity, seasonal activity, ecology, biomass, and demographics of the wild snake community at NCZ [26]. Biologists from the NCNHP assist with monitoring target species and provide expertise on managing for them. NCZ also facilitates research by universities by providing access to natural areas. Some completed projects include a study of longleaf pine (*Pinus palustris*) growth and morphology [27] and a study on the geographical distribution of lichens [28]. NCZ also participates in the Greater Uwharrie Conservation Partnership, which is made up of government agencies and private organizations that work together for the long-term conservation and enhancement of biological diversity and ecosystem sustainability in the Southern Central Piedmont region of North Carolina. The partnership meets regularly to share information and collaborate on projects and grants. By working together, partners increase the efficiency and effectiveness of conservation work in the region.

4. Collaborative Conservation of Declining Amphibians

4.1. *Overview*

Although the North Carolina Zoo works to protect many taxonomic groups, there is a strong focus on the conservation of declining amphibians in North Carolina. Globally, amphibians are more threatened, and are declining more rapidly, than either birds or mammals, with over one-third of all species threatened with extinction [29]. North Carolina is a hotspot for amphibian diversity, with nearly 100 species, including more salamanders than any other region in the world [30], largely because of its diversity of habitats that range from mountain streams to coastal swamps. Unfortunately, many of North Carolina's amphibian species are in decline, mainly due to habitat loss and degradation.

NCZ works on collaborative projects with the North Carolina Wildlife Resources Commission (NCWRC) to bolster populations and improve habitats for target species (Figure 3), including the state-endangered gopher frog (*Lithobates capito*) and the eastern hellbender (*Cryptobranchus alleganiensis*), a species of special concern in North Carolina. Gopher frogs are specialists of the longleaf pine ecosystem, which has disappeared across its range in the southeastern USA [25]. Gopher frogs have declined with their habitat, and the number of populations in North Carolina has decreased from more than 50 to fewer than eight remaining in fragmented areas in southern and southeastern North Carolina [31]. Hellbenders are large, aquatic salamanders found in cool, clean, mountain streams in North Carolina that are threatened by degraded water quality and sedimentation that disrupts microhabitats [32].

Figure 3. North Carolina Zoo works on collaborative projects with the North Carolina Wildlife Resources Commission (NCWRC) to protect declining amphibians. (**A**) Here, NCWRC technician Mike Martin releases a gopher frog that was headstarted at the Zoo into a stump hole, and (**B**) Zoo and NCWRC staff use a hot water pressure washer system to control invasive red imported fire ant colonies to improve habitat for gopher frogs and other species.

Although this section focuses on collaborations for two target amphibian species, NCZ staff also participate in other herpetofaunal surveys across the state with the NCWRC and North Carolina State Parks (NCSP), and they are involved with the North Carolina Partners in Amphibian and Reptile Conservation (NCPARC), which is the local chapter of the Partners in Amphibian and Reptile Conservation (PARC). Such partnerships have proven successful as most organizations do not have sufficient staffing, resources, or experience to achieve long-term success alone. Experience across disciplines, coupled with regular planning and communication, have been paramount to the development of these efforts.

4.2. Population Augmentation

Translocations, which are the intentional movement of individuals or populations across landscapes, have played important roles in the conservation of many species, including amphibians [33–35]. One increasingly used translocation action is population augmentation, which occurs when a declining population is supplemented with individuals from another population [36]. The goal of population augmentation is to build a sustainable population that is resistant to demographic or environmental stochasticity [37,38]. Such population augmentation efforts are long-term endeavors due to numerous abiotic and biotic factors associated with survival and reproduction.

In 2016, NCZ began a collaboration with the NCWRC to augment the declining gopher frog population in the Sandhills region of North Carolina through a headstarting project (Figure 3A). Headstarting is a conservation technique, in which early-stage animals are raised to later life stages in artificial habitats before being released into natural habitats. The goal of headstarting is to increase survival rates from early to later life stages by keeping individuals safe from predators, habitat degradation, and environmental factors. As part of the collaborative effort for headstarting gopher frogs, NCWRC staff collect gopher frog eggs following breeding events and transport them to NCZ. Once the eggs hatch, the tadpoles are reared in outdoor mesocosms until metamorphosis. All newly metamorphosed frogs are measured and marked with Visual Implant Elastomer (VIE) so that they can be identified if they are later observed in the wild. After the frogs are processed, NCZ and NCWRC staff release the frogs into their natural habitat, either near the pond where the eggs were collected (2016–2019) or a nearby restored pond (2020–2022). The husbandry protocol is evaluated each year and adjusted based upon metamorphosis size and percentage of metamorphosed gopher frogs released (Table 1). A similar protocol has also been used on a trial basis to headstart the state-endangered ornate chorus frog (*Pseudacris ornata*) and the eastern tiger salamander (*Ambystoma tigrinum*).

Table 1. Summary of gopher frog headstarting efforts at NCZ from 2016 through 2022. This includes the number of wild egg masses (all collected in early spring, except for 2018, which included fall and spring collections), total number of recently metamorphosed frogs released, the percentage of frogs that survived to metamorphosis and were released, and mean snout-vent length (SVL; not measured in 2016) and mean mass of released frogs. Because of a lack of wild breeding in 2022, all headstarted and released frogs that year were the result of assisted reproductive methods used with frogs housed in NCZ's collection.

Year	Egg Masses	Frogs Released	% Frogs Released	Mean SVL (mm)	Mean Mass (g)
2016	34	266	35	NA	5.7
2017	9	156	78	39.8	7.9
2018 (spring)	3	41	42	38.8	7.4
2018 (fall)	22	170	85	38.2	8.0
2019	19	162	81	37.0	7.2
2020	23	298	75	37.4	8.6
2021	16	459	70 *	34.4	6.2
2022	NA	113	91	34.4	6.4

* Excludes 50 non-releasable frogs removed from the project due to abnormalities.

In 2022, weather patterns hindered many of the winter breeding amphibians in North Carolina and there were no documented reproductive events by gopher frogs in the state. Because there were no eggs to headstart, NCZ and visiting partners from Mississippi State University used assisted reproduction techniques to successfully breed adult gopher frogs in NCZ's managed collection. Although this protocol has been used for gopher frogs and other species elsewhere, it was the first time that in vitro fertilization was used successfully to produce fertile gopher frog eggs in North Carolina. These eggs were raised at NCZ and resulted in the release of over 113 gopher frogs into the Sandhills that year, contributing to 1665 gopher frogs released at two ponds during the seven years of the program (Table 1).

Our monitoring efforts for gopher frogs focus primarily on acoustic monitoring and egg mass surveys because gopher frogs are secretive animals that spend more of their lives underground and are rarely seen outside of the breeding season when they migrate to ponds [39,40]. In the extremely rare event that a gopher frog is encountered outside a breeding pond, it can be visually examined for VIE. During the first four years of the headstarting program, frogs were released at the source pond in which the eggs were collected. Acoustic monitoring and egg mass surveys were conducted throughout the breeding season, and a subset of released frogs were also monitored using radiotelemetry to examine their patterns of movement, behavior, and survival. Many tracked frogs appeared to have been preyed on by red imported fire ants (*Solenopsis invicta*), an introduced species that occurs in high densities at this site. Although survival of juvenile gopher frogs is expected to be low (e.g., 12.5% during the first month; [41]), the threat posed by fire ants led to the development of a plan by NCZ and the NCWRC to control fire ants around this pond (see "Habitat Management" Section below).

Starting in 2020, we began releasing all frogs at a different pond that was recently restored by the NCWRC, but had no records of gopher frog breeding. Acoustic monitoring and egg mass surveys have been conducted throughout each breeding season since releases began. Gopher frog breeding has not yet been recorded at the restored pond, but this is not unexpected because gopher frogs can take several years to mature and only breed under specific weather conditions that do not occur every year. For example, no documented gopher frog breeding occurred anywhere in North Carolina during the spring 2022 breeding season. In collaboration with the NCWRC, our goal is to produce a sustainable population at the restored pond, but it may take many years of headstarting and monitoring to reach this goal due to numerous biotic and abiotic factors associated with survival and reproduction. Without appropriate seasonal rainfall to fill breeding pools and trigger breeding events, frogs may not be encountered for years.

4.3. Habitat Management

For translocations to be successful in building sustainable populations, any threats that caused their decline or could lead to further decline must be addressed [42]. For gopher frogs in the Sandhills population, a serious threat to their survival is invasive red imported fire ants, which can directly kill gopher frogs and other species with their potent venom. There are very high densities of fire ant colonies surrounding the main breeding pond, and many interactions between frogs and ants have been observed, including high predation rates of juvenile gopher frogs monitored using radiotelemetry. To mitigate this severe threat, NCZ and the NCWRC jointly developed a management plan to control fire ants and support gopher frog augmentation efforts. Following the methods of Tschinkel and King [43], the team used a large (1136 L) hot water pressure washer system to treat fire ant colonies (Figure 3B). Since 2020, more than 1200 colonies have been treated in this ongoing project. The treatments have been successful in eliminating some ant colonies; however, the success in reducing fire ant densities on the landscape is under evaluation. Overall, this collaborative project aims to benefit a genetically valuable population of gopher frogs, and the management implications will be applicable throughout the southeastern USA, where red imported fire ants negatively impact populations of many species of amphibians and reptiles.

A second collaborative project among NCZ, the NCWRC, and NCSP aims to benefit hellbender populations by improving habitat through the installation of artificial nest boxes. These boxes are designed to mimic rock crevices and provide nesting habitat and shelter in rivers lacking sufficient breeding sites. There are multiple designs used throughout the range of both the eastern and Ozark hellbender subspecies. However, despite nest boxes being extremely useful in Virginia, Ohio, and other parts of the hellbender's range, those used in North Carolina have been considered less successful because of little to no occupancy and lack of breeding success. NCZ is working with the NCWRC and NCSP to test hellbender nest box occupancy along stretches of the New River, where hellbenders have been monitored by NCSP staff and volunteers for over 10 years. Although this population seems to be stable, numerous populations of hellbenders in North Carolina have declined since 2010, by as much as 50% (John Groves, NCZ retired, and Lori Williams, the NCWRC, personal communication). The goal is to develop successful methods for installing and managing hellbender nest boxes in North Carolina by monitoring these boxes in sites with known hellbender populations so that these methods can be applied to other sites to benefit hellbender populations.

To date, NCZ has installed more than 40 nest boxes at two sites within the known range of hellbenders in North Carolina. Efforts at the first site were not effective after three years, primarily due to stochastic weather events and heavy river flow. Due to these factors, the boxes were moved to the New River State Park in 2020 as part of this new collaboration. While we have not documented occupancy at either site, the flow of the New River has been less impactful on box placement, and multiple crayfish and fish species have been documented using the boxes. We collect standardized data during each monthly survey during warmer months, which includes recording river width, flow rate, depth, and water quality, so we can make comparisons with other sites to examine nest box efficacy.

5. Wildlife Rehabilitation

Wildlife rehabilitation is the care of sick, injured, orphaned, and displaced animals for return to their natural habitat. Rehabilitation is provided to animals that have been orphaned due to human intervention or developmental abnormalities; sick animals exposed to toxins or carrying parasites or pathogens; injured animals that have experienced trauma due to collision (e.g., with windows and vehicles), entrapment/entanglement, or injury by another animal or human; and animals that have been displaced through habitat loss (e.g., tree removal, construction, and natural disturbances) [44–46]. Because cases are usually related to anthropogenic activity, the purpose of wildlife rehabilitation is to offset human impact and to mitigate human–wildlife conflict [47]. The value of wildlife rehabilitation is multi-fold, from ensuring native animals are cared for by experienced and equipped rehabilitators, to educating the public about wildlife, to engaging the community in assisting with the care of wildlife. Wildlife rehabilitation has conservation and scientific purposes as well. For example, rehabilitation has been instrumental in disease surveillance [48–51]) and providing a better understanding of the natural history and needs of species while in human care. In addition, by providing people with close contact with wildlife, rehabilitation can increase knowledge and respect of local wildlife and the threats they face, which can contribute to biodiversity conservation.

North Carolina Zoo's Wildlife Rehabilitation Center (Center) opened in August 2001 as a community resource for assisting with local wildlife in need of veterinary and rehabilitative care (Figure 4). The Center's goal is to offset human impact and inspire and educate people about native wildlife. The Center provides free, professional veterinary and rehabilitation services to sick, injured, and orphaned native North Carolina animals found in the wild for the sole purpose of returning them to the wild in a condition that will optimize their chances of survival after release. The Center performs euthanasia when injuries are too severe for animals to live a quality life. The Center admits 800–1000 animals per year, representing over 100 species. Most species admitted are common because these are most likely to be encountered by the public. These include birds common in residential and urban areas (e.g., songbirds, woodpeckers, and doves), eastern gray squirrels (*Sciurus carolinensis*), eastern cottontails (*Sylvilagus floridanus*), Virginia opossums (*Didelphis virginiana*), and eastern box turtles (*Terrapene carolina*; Figure 4A). In addition, the Center has rehabilitated uncommon species, such as bald eagles (*Haliaeetus leucocephalus*), and protected species, including wood storks (*Mycteria americana*), piping plovers (*Charadrius melodus*), timber rattlesnakes (*Crotalus horridus*), and diamond-back terrapins (*Malaclemys terrapin*). Rehabilitation can play an important role in species conservation, especially for long-lived species with population declines that are difficult to detect, such as eastern box turtles that take up to 10 years to reach sexual maturity and can live for 50–100 years or more [52]. Most frequently, animals are admitted due to vehicular collisions, injuries caused by domestic cats, window collisions, and unnecessary intervention of presumed orphans. The Center is uniquely qualified to care for species with specialized needs as it has the resources of NCZ, including licensed rehabilitators, board-certified veterinarians, expertise of NCZ staff, and facilities that most independent wildlife caretakers and clinics do not have at their disposal. This includes a large variety of enclosures, shift cages, aviaries, and outdoor "wilding" or "pre-release" conditioning habitats that meet or exceed standard guidelines [53].

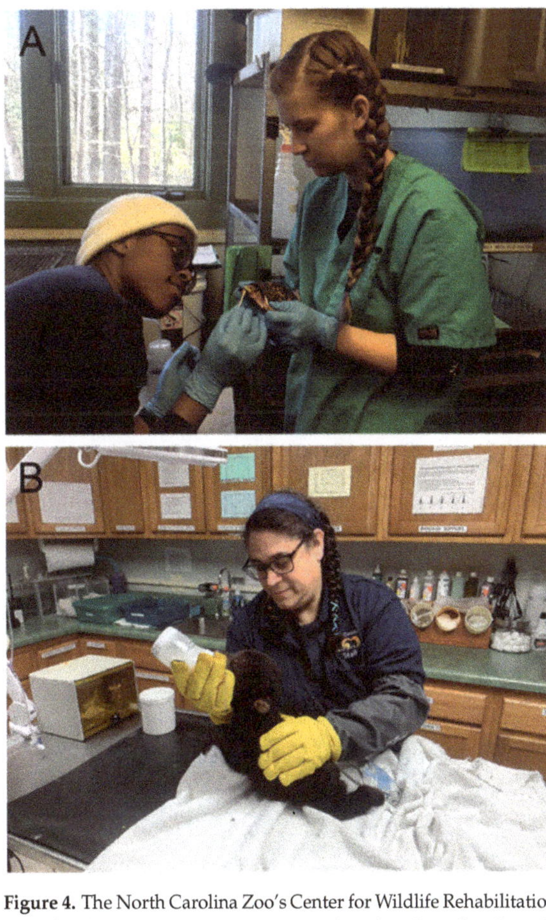

Figure 4. The North Carolina Zoo's Center for Wildlife Rehabilitation provides professional veterinary and rehabilitation services to sick, injured, and orphaned native wildlife, such as (**A**) this eastern box turtle being treated by a veterinary student and intern, and (**B**) this orphaned black bear cub reared through a partnership with the North Carolina Wildlife Resources Commission.

Initially, the Center was staffed by volunteers, with care overseen by NCZ's veterinary staff, but as the caseload increased as local residents became aware of the Center and services provided, there was a need for onsite staff. Consequently, a veterinary technician was hired to manage Center operations in 2007. Since then, the Center has grown to include two full-time staff members, three seasonal staff members, a well-established and reputable internship program, a stable volunteer program, and visiting veterinary students from North Carolina State University's Cummings School of Veterinary Medicine. The Center trains interns, volunteers, and veterinary students in respectful and responsible wildlife rehabilitation methods using the most up-to-date and science-based practices. Since its inception, the Center has hosted over 60 interns from more than 25 states. The internship program includes a weekly curriculum that is accredited by the International Wildlife Rehabilitation Council and hands-on opportunities working directly with wildlife with training and supervision from Center staff.

In addition to providing rehabilitative care, the Center also participates in research. Post-release survival studies have been conducted on rehabilitated birds since 2012 that are banded with metal bands issued by the United States Geological Survey and approved sequences of color bands to make sure each bird is individually recognizable from a distance. This study focuses on five species: blue jays (*Cyanocitta cristata*), Carolina wrens (*Thryothorus ludovicianus*), mourning doves (*Zenaida macroura*), northern cardinals (*Cardinalis cardinalis*), and red-bellied woodpeckers (*Melanerpes carolinus*). The Center initially partnered with Guilford College to initiate the study and later acquired permits to continue the study independently [54]. As part of this ongoing study, birds were admitted to the Center at fledging stage or younger and cared for until they were released at the appropriate stage. All birds were released outside the Center by placing them in a soft-sided enclosure with perches that was hung at the release site for a minimum of 30 min and then unzipped, allowing the birds to fly out on their own. Birds were resighted opportunistically and evaluated according to an ethogram to assess whether released birds were behaving similar to their wild, parent-reared counterparts or behaving inappropriately due to their time in human care. Between 2012 and 2022, 320 birds were banded as part of this ongoing study evaluating the success of hand-reared birds. Of these birds, 42 (13.1%) were resighted at least once and up to 327 times. Birds survived up to 6.3 years post-release and the average number of days to the last resighting was 243 days. All resighted birds demonstrated normal wild behavior that was indistinguishable from their wild counterparts.

As part of a collaboration with the NCWRC that began in 2015, the Center rears orphaned black bear (*Ursus americanus*) cubs (Figure 4B) and assists biologists with fitting the cubs with GPS collars so they can be tracked after they are released to study their movement and behavior. The Center is one of two facilities in the state approved by the NCWRC to rear black bear cubs because the Center can provide a large, safe space for them to grow and is committed to rearing cubs in a hands-off manner, as much as possible. Preliminary data from this ongoing study suggest that the behavior and survival of released bears appears similar to those of wild-reared bears and the majority have not become nuisance bears.

The need for better rehabilitation programs for North American river otters (*Lontra canadensis*) has also led to a partnership with the North Carolina Aquariums. River otters are intelligent and inquisitive, and they can readily habituate to caregivers during hand-rearing. As they age, they must have access to deep water pools to learn to swim and catch fish and disassociate with humans. The North Carolina Aquariums often receive calls about orphaned otter pups, and they will assess whether they can be reunited, need to be rehabbed, or need to be placed under permanent human care. If rehabilitation of otter pups is warranted, the Center will step in to assist with rearing.

The Center has assisted thousands of North Carolina residents with wildlife in need by advising callers with concerns about wildlife, admitting and caring for wildlife in need of veterinary or rehabilitative care, training interns and volunteers about wildlife husbandry and medicine, and educating other rehabilitators through conference presentations and networking. When animals have a poor prognosis for recovery, the Center provides euthanasia and the carcasses can be used for educational purposes, either for scientific study or training individuals in veterinary care. Through numerous partnerships, NCZ's Wildlife Rehabilitation Center has become a vital community resource in engaging people to protect wildlife and wild places.

6. Outreach and Sustainability

6.1. Connecting People with Nature

Hiking trails connect people with nature and improve their physical and mental health, while benefitting the environment and local economies [22–24,55]. North Carolina Zoo has long embraced the idea that trails are one of the best ways to introduce people to nature. Because of this mindset, NCZ openly advocates for trails, both on NCZ property and in the region. Since 2019, NCZ has been a partner in the Randolph County Trails Advisory Commission. This partnership includes representatives from nine municipalities across the county, as well as other representatives that have an interest in promoting outdoor activities and protecting natural and historical areas. Together, this group advocates for trail expansion and protection of natural and cultural resources by sharing information, applying for grants to acquire property and build trails, and planning and completing these projects.

NCZ builds and maintains sustainable hiking trails that protect sensitive plants and animals while providing opportunities for hikers to explore natural features on NCZ's main property (Purgatory Mountain) and Ridges Mountain Preserve. NCZ currently maintains approximately eight miles of trails and plans to build additional trails. Trails on Purgatory Mountain begin at NCZ's North America region parking lot and are open at no charge anytime NCZ gates are open, whereas trails at Ridges Mountain Preserve require hikers to notify NCZ before visiting. NCZ trails systems are among the most significant in the area, in terms of distances, interesting features, and usage by hikers. The Purgatory Mountain Trail System (8 km) is the second largest, behind the Birkhead Mountains Wilderness (24 km; Figure 1), whereas most other trails in the region are shorter in distance (<3 km) and tend to be in more urban areas. We have observed that the Purgatory Mountain trails are used daily during favorable weather by multiple groups of visitors, as well as NCZ staff enjoying a hike on their lunch break or before or after their shift. Most non-staff users are local, but others have traveled from farther locations within North Carolina or other states. The NCZ is often the main destination for these guests, but they extend their stay to hike on the trails while they are in the area. The internal Trail Team leads planning, construction, and maintenance of all trails, and assistance with trail work is provided through partnerships with local schools and other volunteer groups (Figure 5A). Partnerships with the Phi Theta Kappa Chapter at Randolph Community College and Future Farmers of America at Asheboro High School Zoo School have been especially important for completing trail work and engaging students during the COVID-19 pandemic when other service opportunities were suspended. NCZ naturalists lead regular nature hikes for public and private groups along the trails to explore the biodiversity and unique geologic features on the land.

Figure 5. North Carolina Zoo works on local outreach and sustainability projects with community partners to reduce impacts on natural resources and inspire guests and community members to get involved in conservation. This includes partnering with local schools to (**A**) build sustainable hiking trails on zoo-owned land and (**B**) plant trees on school grounds.

6.2. Combating Habitat Loss and Degradation

Habitat loss and degradation are the largest threats to biodiversity globally [17,18], and NCZ combats these threats by increasing habitat quality for native species on NCZ grounds and in the local community. Pollinators are important components of ecosystems [56], and NCZ promotes their conservation as a program partner in the AZA SAFE (Saving Animals From Extinction) North American Monarch program. One of the actions NCZ takes to enhance habitat for monarchs and other pollinators is planting Monarch Waystations and Pollination Stations. Monarch butterflies are threatened by habitat loss at overwintering grounds in Mexico and throughout breeding areas in the United States and Canada [57]. The Monarch Waystation program, led by the organization Monarch Watch, promotes planting gardens that contain milkweeds and nectar plants to support larval and adult monarchs. NCZ maintains five Monarch Waystations on grounds that support monarchs and other pollinators and provide interpretive signs to educate NCZ guests about monarchs and their habitat requirements. We have observed monarch eggs and caterpillars on common milkweed (*Asclepias syriaca*) and swamp milkweed (*Asclepias incarnata*) in the gardens. Adult monarch butterflies and other pollinators, such as bees, moths, and other butterflies have also been observed using nectar plants in the gardens. These include bumblebees (*Bombus* spp.), carpenter bees (*Xylocopa* spp.), mason bees (*Osmia* spp.), hummingbird clearwings (*Hemaris thysbe*), and eastern tiger swallowtails (*Papilio glaucus*). NCZ has also partnered with five local schools and the Randolph County Public Library to support pollinators and educate community members by creating "Pollination Stations" at their locations. These container gardens are planted in upcycled containers provided by NCZ.

NCZ also supports songbird conservation as a program partner in the AZA SAFE North American Songbirds program. The goal of this program is to reduce threats to North American songbirds and secure sustainable wild populations of these species throughout their ranges by harnessing the collective strengths of zoos, aquariums, and partners through supporting education and on-the-ground conservation activities. One way that NCZ participates in this program is by keeping windows and other glass surfaces safe for birds. Up to one billion birds die from window collisions in the USA each year [58], making this the second largest threat to birds in the country (behind domestic cats). Research on bird-glass collisions has led to the development of several different types of commercially available products that reduce glass reflection and transparency and successfully reduce collisions [59–61]. NCZ has installed UV-reflective glass and patterned films on windows and other glass surfaces that are part of animal exhibits and viewing areas. Interpretive signs below these glass treatments educate NCZ guests about the threats of windows to birds and what they can do to help at home. NCZ is closely monitoring the effectiveness of these glass treatments. For example, when Feather Friendly® window markers were installed on a building containing many large, mirrored windows in 2021, bird collisions declined by 68% (from 38 to 12 strikes) during the year following mitigation when compared with the previous year. Another study observed a similar reduction in collisions (71%) after installing the same product [60].

A major environmental challenge is litter pollution, which can harm wildlife when they ingest it directly or indirectly, become trapped in it, or become drawn to roadsides and other dangerous areas where litter accumulates [62–64]. One way that NCZ combats this issue is by participating in the Adopt-A-Highway program, managed by the North Carolina Department of Transportation. Through this program, NCZ has conducted 5–6 cleanups per year for the last 25 years, removing thousands of pounds of trash from a two-mile stretch of roadway. NCZ also keeps cell phones and other small electronics out of natural areas and landfills by collecting them for recycling through the Gorillas on the Line program. This global program aims to increase recycling of cell phones and small electronics to protect habitats for gorillas and other wildlife in central Africa, where materials are mined. Furthermore, NCZ composts 2000 tons of waste annually, including animal manure and plant waste, in addition to food scraps and compostable plates, bowls, and cutlery from the onsite restaurants. The finished compost is used in NCZ's horticulture operations.

6.3. Combating Climate Change

Climate change is a major threat to biodiversity, but accurate predictions and effective solutions are difficult to develop [65,66]. NCZ works with partners to combat climate change at the local scale by planting trees, promoting electric vehicles, and educating and empowering people to take action against climate change. Though a partnership with Polar Bears International as an Arctic Ambassador Center, NCZ works collaboratively on research, education, and action programs that address climate change. Some of the programs implemented by NCZ as part of this partnership involve planting trees to increase sinks for atmospheric carbon. Over the last 12 years, NCZ has planted over 600 trees at schools, parks, and other public lands in the community, and distributed thousands of seedlings as part of Arbor Day and Earth Day celebrations (Figure 5B). Using a conservative estimate (10 tons of carbon dioxide sequestered per hectare by 1000 trees; [67]), we estimate that an average tree absorbs an average of 10 kg of carbon dioxide per year, and therefore, that the 600 trees we have planted absorb 6000 kg of carbon dioxide each year. In addition to their value in storing carbon, some of the trees were planted to absorb rainwater and mitigate erosion or to provide shade in heat islands, thereby providing other benefits to the environment and community.

NCZ installed a grid-connected photovoltaic solar array in 2008 called "Solar Pointe" as another initiative to combat climate change. Each of the three picnic shelters at Solar Pointe supports 297 square meters of photovoltaic panels, and together they can generate a maximum of 104 kilowatts of power per hour. These panels supply energy to the local power grid through partnerships with Carolina Solar Electric Company and the Randolph Electric Membership Cooperative (REMC). Solar Pointe is not capable of generating enough energy to fully operate NCZ (1.8 megawatts used per hour at peak operation), but the supply helps reduce the local demand for energy and avoid potential issues during extreme weather when demand could exceed supply. In addition, for each hour that Solar Pointe is generating energy at maximum capacity, 0.074 tons of greenhouse gas emissions are avoided, which is equivalent to burning 37 kg of coal. With an estimated 300 days of clear skies annually over fourteen years and a daily maximum generating time of four hours per day, Solar Pointe has avoided the use of 684 tons of coal. This is comparable to the energy required to heat and cool 17 residential homes for one year.

NCZ also reduces reliance upon petroleum products by using and promoting electric vehicles. There are currently 13 electric vehicles in NCZ's vehicle fleet, and two electric buses have been ordered with grant funding from the North Carolina Department of Environmental Quality (NCDEQ). These vehicles produce no direct emissions, but they are charged by connecting to the local power grid. REMC receives 8% of energy from renewable sources (e.g., solar and wind), 2% from hydroelectric installations, and 54% from a nuclear power plant (Michael Trent, REMC, personal communication). Because nuclear power reactors do not produce direct carbon dioxide emissions, they are largely carbon neutral, although the overall impact of nuclear power on the environment is controversial. Because NCZ's electric vehicles produce no direct emissions and a portion of the energy used to power them is carbon neutral, they operate more efficiently and with less air pollution than their gasoline or diesel counterparts. NCZ also provides electric vehicle chargers as a free service to guests to encourage electric vehicle use. This infrastructure was funded by a grant from NCDEQ. Annually, in collaboration with the Randolph Electric Membership Cooperative, NCZ hosts the annual Kickoff to National Drive Electric Week event. This event showcases the clean-air and cost-saving benefits of driving all-electric vehicles and plug-in hybrid vehicles. Through these initiatives and others, NCZ is committed to sustainability by reducing its impact and raising awareness about green practices.

7. Recommendations for Zoos

NCZ's regional conservation programs have evolved over time and continue to develop in response to outcomes and conservation needs. NCZ is unique in some ways because it is surrounded by natural areas with high biodiversity and is supported by the state and a non-profit partner. However, there are many ways that other zoos, including urban zoos or smaller zoos with fewer resources, can contribute to local conservation efforts in meaningful ways. Many local initiatives are inexpensive or only require participation by staff or volunteers, since extensive travel is not required, unlike many international conservation efforts. Although many zoos do not own conservation lands, they can create wildlife habitat on their grounds by planting native vegetation that benefits wildlife and pollinators, and avoiding planting non-native species that may cause ecological harm. Zoos can also partner with local state, county, or city parks by providing leadership and stewardship. For example, zoos can provide staff and volunteers to assist parks with planting trees, removing invasive plants, building trails, or cleaning up litter. Zoos could also partner with parks to lead public nature hikes or contribute to biodiversity monitoring. For example, Memphis Zoo partners with the Overton Park Conservancy in conducting stewardship, outreach, and research activities in the Old Forest State Natural Area, adjacent to the Memphis Zoo.

Zoos can also support regional conservation by participating in large-scale community science projects such as NestWatch or FrogWatch and contribute observations on biodiversity for use in science through platforms such as eBird and iNaturalist. If zoos are interested in playing a larger role in conservation efforts for local imperiled species, we recommend connecting with the state wildlife agency to discuss possible partnerships. Zoos often play an important role in headstarting and captive breeding programs because many have facilities and expertise for caring for animals before they are released into natural habitats. Examples include Woodland Park Zoo and Oregon Zoo's headstarting program for western pond turtles (*Actinemys marmorata*), and the captive breeding and release program for the eastern indigo snake (*Drymarchon couperi*) at the Orianne Center for Indigo Conservation, which is operated by the Central Florida Zoo and Botanical Gardens. Finally, zoos can provide important conservation messaging through signage and social media to educate others about wildlife and inspire actions they can take to conserve wildlife and their habitats. The AZA SAFE programs (e.g., SAFE North American Songbird and SAFE North American Monarch) provide useful resources for messaging, as well as recommendations for conservation initiatives that can be implemented at participating zoos.

8. Conclusions

Protecting biodiversity depends on the contributions of committed conservation organizations working together to solve complex challenges. The North Carolina Zoo works with partners across the state on projects that aim to protect local species and their habitats. NCZ plays an important role in land conservation by protecting and managing more than 800 ha of undeveloped land, and multiple partnerships have been key in identifying, acquiring, and managing these ecologically significant lands. NCZ also works on collaborative projects with the NCWRC to protect declining amphibians, including supporting gopher frog populations through headstarting and habitat management, and installing artificial nest boxes to provide nesting habitat and shelter for hellbenders. In partnership with the NCWRC and North Carolina Aquariums, NCZ's Wildlife Rehabilitation Center cares for wildlife in need of veterinary or rehabilitative care, with support from trained interns, veterinary students, and volunteers. NCZ also collaborates with numerous community partners on numerous outreach and sustainability initiatives, including building hiking trails, planting pollinator gardens and trees, keeping windows safe for birds, removing litter, composting, and installing solar panels and electric vehicle chargers.

Zoos are becoming leaders in conservation, research, and education as their role evolves over time. To address growing biodiversity losses and human impacts on the environment, collaborative efforts are needed to provide solutions that are effective, innovative, and long-lasting. As demonstrated through the conservation programs reviewed here, zoos can develop successful partnerships with local and state government agencies, academic institutions, non-profit organizations, other zoos and aquariums, schools, libraries, and businesses. Zoos can benefit these partnerships by contributing crucial resources and expertise that are needed to protect natural communities and the benefits they provide to humans. Through partnerships that aim to reduce their ecological footprints, zoos can benefit local and global ecosystems and play a lead role in setting new sustainability standards and serving as examples for other attractions and organizations.

Author Contributions: Conceptualization, E.A.R., H.B., R.W.L., C.J.S. and D.S.; data curation, E.A.R., H.B., R.W.L., C.J.S. and D.S.; formal analysis, E.A.R., H.B., R.W.L., C.J.S. and D.S.; investigation, E.A.R., H.B., R.W.L., C.J.S. and D.S.; methodology, E.A.R., H.B., R.W.L., C.J.S. and D.S.; writing—original draft preparation, E.A.R., H.B., R.W.L. and D.S.; writing—review and editing, E.A.R., H.B., R.W.L., C.J.S. and D.S.; visualization, E.A.R. All authors have read and agreed to the published version of the manuscript.

Funding: This research received no external funding.

Institutional Review Board Statement: Not applicable.

Data Availability Statement: The data presented in this study will be made openly available in FigShare upon acceptance of the manuscript.

Acknowledgments: We thank our many partners across North Carolina for their support and dedication to conservation, and we are grateful to our guests, members, donors, and the North Carolina Zoological Society for making this important work possible. We thank Nell Allen and Jeff Hall for providing helpful feedback on the manuscript.

Conflicts of Interest: The authors declare no conflict of interest.

References

1. Barnoski, A.D.; Matzke, N.; Tomiya, S.; Wogan, G.O.U.; Swatz, B.; Quental, T.B.; Marshall, C.; McGuire, J.L.; Lindsey, E.L.; Maquire, K.C.; et al. Has the Earth's sixth mass extinction already arrived? *Nature* **2011**, *471*, 51–57. [CrossRef]
2. Dirzo, R.; Young, H.S.; Galetti, M.; Ceballos, G.; Isaac, N.J.B.; Collen, B. Defaunation in the Anthropocene. *Science* **2014**, *345*, 401–406. [CrossRef] [PubMed]
3. Pimm, S.L.; Jenkins, C.N.; Abell, R.; Brooks, T.M.; Gittleman, J.L.; Joppa, L.N.; Raven, P.H.; Roberts, C.M.; Sexton, J.O. The biodiversity of species and their rates of extinction, distribution, and protection. *Science* **2014**, *344*, 1246752. [CrossRef] [PubMed]
4. Ceballos, G.; Ehrich, P.R.; Dirzo, R. Biological annihilation via the ongoing sixth mass extinction signaled by vertebrate population losses and declines. *Proc. Natl. Acad. Sci. USA* **2017**, *114*, E6089–E6096. [CrossRef] [PubMed]
5. Wilkins, K.; Pejchar, L.; Carroll, S.L.; Jones, M.S.; Walker, S.E.; Shinbrot, X.A.; Huayhuaca, C.; Fernández-Giménez, M.E.; Reid, R.S. Collaborative conservation in the United States: A review of motivations, goals, and outcomes. *Biol. Conserv.* **2021**, *259*, 109165. [CrossRef]
6. Raschke, A.B.; Pegram, K.V.; Melkonoff, N.A.; Davis, J.; Blackwell, S.A. Collaborative conservation by botanical gardens: Unique opportunities for local to global impacts. *J. Zool. Bot. Gard.* **2022**, *3*, 463–487. [CrossRef]
7. Mallinson, J.J.C. A sustainable future for zoos and their role in wildlife conservation. *Hum. Dimens. Wildl.* **2003**, *8*, 59–63. [CrossRef]
8. Tribe, A.; Booth, R. Assessing the role of zoos in wildlife conservation. *Hum. Dimens. Wildl.* **2003**, *8*, 65–74. [CrossRef]
9. Bayma, T. Rational myth making and environment shaping: The transformation of the zoo. *Sociol. Q.* **2012**, *53*, 116–141. [CrossRef]
10. Association of Zoos and Aquariums. *2021 Annual Report on Conservation and Science Highlights*; Association of Zoos and Aquariums: Silver Spring, MD, USA, 2021.
11. Association of Zoos and Aquariums Visitor Demographics. Available online: https://www.aza.org/partnerships-visitor-demographics?locale=en (accessed on 10 February 2023).
12. Gusset, M.; Dick, G. The global reach of zoos and aquariums in visitor numbers and conservation expenditures. *Zoo Biol.* **2011**, *30*, 566–569. [CrossRef]
13. Braverman, I. Conservation without nature: The trouble with in situ versus ex situ conservation. *Geoforum* **2014**, *51*, 47–57. [CrossRef]
14. North Carolina Zoo. *North Carolina Zoological Park Land Use Plan*; North Carolina Zoo: Asheboro, NC, USA, 2000.
15. Geldmann, J.; Barnes, M.; Coad, L.; Craigie, I.D.; Hockings, M.; Burgess, N.D. Effectiveness of terrestrial protected areas in reducing habitat loss and population declines. *Biol. Conserv.* **2013**, *161*, 230–238. [CrossRef]
16. Gray, C.L.; Hill, S.L.L.; Newbold, T.; Hudson, L.N.; Börger, L.; Contu, S.; Hoskins, A.J.; Ferrier, S.; Purvis, A.; Schartemann, J.P.W. Local biodiversity is higher inside than outside terrestrial protected areas worldwide. *Nat. Commun.* **2016**, *7*, 12306. [CrossRef] [PubMed]
17. Bender, D.J.; Contreras, T.A.; Fahrig, L. Habitat loss and population decline: A meta-analysis of the patch size effect. *Ecology* **1998**, *79*, 517–533. [CrossRef]
18. Fischer, J.; Lindenmayer, D.B. Landscape modification and habitat fragmentation: A synthesis. *Glob. Ecol. Biogeogr.* **2007**, *16*, 265–280. [CrossRef]
19. Devictor, V.; Godet, L.; Julliard, R.; Couvet, D.; Jiguet, F. Can common species benefit from protected areas? *Biol. Conserv.* **2007**, *139*, 29–36. [CrossRef]
20. Lyons, K.G.; Brigham, C.A.; Traut, B.H.; Schwartz, M.W. Rare species and ecosystem functioning. *Conserv. Biol.* **2005**, *19*, 1019–1024. [CrossRef]

21. Winfree, R.; Fox, J.W.; Williams, N.M.; Reilly, J.R.; Cariveau, D.P. Abundance of common species, not species richness, drives delivery of a real-world ecosystem service. *Ecol. Lett.* **2015**, *18*, 626–635. [CrossRef]
22. Lachowycz, K.; Jones, A.P. Towards a better understanding of the relationship between greenspace and health: Development of a theoretical framework. *Landsc. Urban Plan.* **2013**, *118*, 62–69. [CrossRef]
23. Mitten, D.; Overholt, J.R.; Haynes, F.I.; D'Amore, C.C.; Ady, J.C. Hiking: A low-cost accessible intervention to promote health benefits. *Am. J. Lifestyle Med.* **2018**, *2018*, 302–310. [CrossRef]
24. Hoover, K.S. Children in nature: Exploring the relationship between childhood outdoor experience and environmental stewardship. *Environ. Educ. Res.* **2021**, *27*, 894–910. [CrossRef]
25. Jose, S.; Jokela, E.J.; Miller, D.L. (Eds.) *The Longleaf Pine Ecosystem: Ecology, Silviculture, and Restoration*; Springer: New York, NY, USA, 2006.
26. Smith, D.C.; Reynolds, R.G.; Hagen, K. Snakes on a path: Ecology of a North Carolina Piedmont snake community. *Herpetol. Rev.* **2021**, *52*, 473–481.
27. Patterson, T.W.; Cummings, L.W.; Knapp, P.A. Longleaf pine (*Pinus palustris* Mill.) morphology and climate/growth responses along physiographic gradient in North Carolina. *Prof. Geogr.* **2016**, *68*, 238–248. [CrossRef]
28. Perlmutter, G.B.; Plata, E.A. Lichens of Purgatory and Ridges Mountains: Further explorations in the Uwharrie Mountains of North Carolina, USA. In Proceedings of the Seventh International Lichenological Symposium, Bangkok, Thailand, 9–13 January 2012.
29. Stuart, S.N.; Chanson, J.S.; Cox, N.A.; Young, B.E.; Rodrigues, A.S.L.; Fischman, D.L.; Waller, R.W. Status and trends of amphibian declines and extinctions worldwide. *Science* **2004**, *306*, 1783–1786. [CrossRef]
30. Beane, J.C.; Braswell, A.L.; Mitchell, J.C.; Palmer, W.M.; Harrison, J.R., III. *Amphibians and Reptiles of the Carolinas and Virginia*; University of North Carolina Press: Chapel Hill, NC, USA, 2010.
31. North Carolina Wildlife Resources Commission. *Gopher Frog Conservation Plan for North Carolina*; North Carolina Wildlife Resources Commission: Raleigh, NC, USA, 2020.
32. Pugh, M.W.; Hutchins, M.; Madritch, M.; Siefferman, L.; Gangloff, M.M. Land-use and local physical and chemical habitat parameters predict site occupancy by hellbender salamanders. *Hydrobiologia* **2016**, *770*, 105–116. [CrossRef]
33. Griffith, B.; Scott, J.M.; Carpenter, J.W.; Reed, C. Translocation as a species conservation tool: Status and strategy. *Science* **1989**, *245*, 477–480. [CrossRef]
34. Fischer, J.; Lindenmayer, D.B. An assessment of the published results of animal relocations. *Biol. Conserv.* **2000**, *96*, 1–11. [CrossRef]
35. Germano, J.M.; Bishop, P.J. Suitability of amphibians and reptiles for translocation. *Conserv. Biol.* **2009**, *23*, 7–15. [CrossRef] [PubMed]
36. Byrne, P.G.; Silla, A.J. An experimental test of the genetic consequences of population augmentation in an amphibian. *Conserv. Sci. Pract.* **2020**, *2*, e194. [CrossRef]
37. Kronenberger, J.A.; Gerberich, J.C.; Fitzpatrick, S.W.; Broder, E.D.; Angeloni, L.M.; Funk, W.C. An experimental test of alternative population augmentation scenarios. *Conserv. Biol.* **2018**, *32*, 838–848. [CrossRef] [PubMed]
38. Weeks, A.R.; Sgro, C.M.; Young, A.G.; Frankham, R.; Mitchell, N.J.; Miller, K.A.; Byrne, M.; Coates, D.J.; Eldridge, M.D.B.; Sunnucks, P.; et al. Assessing the benefits and risks of translocations in changing environments: A genetic perspective. *Evol. Appl.* **2011**, *4*, 709–725. [CrossRef]
39. Roznik, E.A.; Johnson, S.A.; Greenberg, C.H.; Tanner, G.W. Terrestrial movements and habitat use of gopher frogs in longleaf pine forests: A comparative study of juveniles and adults. *For. Ecol. Manag.* **2009**, *259*, 187–194. [CrossRef]
40. Humphries, W.J.; Sisson, M.A. Long distance migrations, landscape use, and vulnerability to prescribed fire of the gopher frog (*Lithobates capito*). *J. Herpetol.* **2012**, *46*, 665–670. [CrossRef] [PubMed]
41. Roznik, E.A.; Johnson, S.A. Burrow use and survival of newly metamorphosed gopher frogs (*Rana capito*). *J. Herpetol.* **2009**, *43*, 431–437. [CrossRef]
42. Semlitsch, R.D. Critical elements for biologically based recovery plans of aquatic-breeding amphibians. *Conserv. Biol.* **2002**, *16*, 619–629. [CrossRef]
43. Tschinkel, W.R.; King, J.R. Targeted removal of ant colonies in ecological experiments, using hot water. *J. Insect Sci.* **2007**, *7*, 41. [CrossRef]
44. Molina-López, R.A.; Casal, J.; Darwich, L. Causes of morbidity in wild raptor populations admitted at a wildlife rehabilitation centre in Spain from 1995–2007: A long term retrospective study. *PLoS ONE* **2011**, *6*, e24603. [CrossRef]
45. Demezas, K.G.; Robinson, W.D. Characterizing the influence of domestic cats on birds with wildlife rehabilitation center data. *Diversity* **2021**, *13*, 322. [CrossRef]
46. Hanson, M.; Hollingshead, N.; Schuler, K.; Siemer, W.F.; Martin, P.; Bunting, E.M. Species, causes, and outcomes of wildlife rehabilitation in New York State. *PLoS ONE* **2021**, *16*, e0257675. [CrossRef]
47. Long, R.B.; Krumlauf, K.; Young, A.M. Characterizing trends in human-wildlife conflicts in the American Midwest using wildlife rehabilitation records. *PLoS ONE* **2020**, *15*, e0238805. [CrossRef]
48. Trocini, S.; Paciono, C.; Warrem, K.; Butcher, J.; Robertson, I. Wildlife disease passive surveillance: The potential role of wildlife rehabilitation centres. In Proceedings of the National Wildlife Rehabilitation Conference, Canberra, Australia, 22–24 July 2008.

49. Randall, N.J.; Blitvich, B.J.; Blanchong, J.A. Efficacy of wildlife rehabilitation centers in surveillance and monitoring of pathogen activity: A case study with West Nile virus. *J. Wildl. Dis.* **2012**, *48*, 646–653. [CrossRef]
50. Yabsley, M.J. The role of wildlife rehabilitation in wildlife disease research and surveillance. In *Medical Management of Wildlife Species: A Guide for Practitioners*; Hernandez, S.M., Barron, H.W., Miller, E.A., Aguilar, R.F., Yabsley, M.J., Eds.; John Wiley & Sons: Hoboken, NJ, USA, 2019; pp. 159–165.
51. Franzen-Klein, D.; Adamovicz, L.; McRuer, D.; Carroll, S.A.; Wellehan, J.F.X.; Allender, M.C. Prevalence of box turtle adenovirus in eastern box turtles (*Terrapene carolina carolina*) presented to a wildlife rehabilitation center in Virginia, USA. *J. Zoo Wildl. Med.* **2020**, *50*, 769–777. [CrossRef] [PubMed]
52. Dodd, K.L., Jr. *North American Box Turtles: A Natural History*; University of Oklahoma Press: Norman, OK, USA, 2001.
53. Miller, E.A.; Schlieps, J. (Eds.) *Standards for Wildlife Rehabilitation*; National Wildlife Rehabilitators Association: Bloomington, MN, USA, 2021.
54. Buckanoff, H.D.; Moseley, L.J. Post-release monitoring of hand-reared songbirds. *J. Wildl. Rehabil.* **2015**, *35*, 7–10.
55. Zaradic, P.A.; Pergams, O.R.W.; Kareiva, P. The impact of nature experience on willingness to support conservation. *PLoS ONE* **2009**, *4*, e7367. [CrossRef] [PubMed]
56. Potts, S.G.; Imperatriz-Fonseca, V.; Ngo, H.T.; Aizen, M.A.; Biesmeijer, J.C.; Breeze, T.D.; Dicks, L.V.; Garibaldi, L.A.; Hill, R.; Settele, J.; et al. Safeguarding pollinators and their values to human well-being. *Nature* **2016**, *540*, 220–229. [CrossRef]
57. Flockhart, D.T.T.; Pichancourt, J.-B.; Norris, D.R.; Martin, T.G. Unravelling the annual cycle in a migratory animal: Breeding-season habitat loss drives population declines of monarch butterflies. *J. Anim. Ecol.* **2016**, *84*, 155–165. [CrossRef]
58. Loss, S.L.; Will, T.; Loss, S.S.; Marra, P.P. Bird–building collisions in the United States: Estimates of annual mortality and species vulnerability. *Condor* **2014**, *116*, 8–23. [CrossRef]
59. Klem, D., Jr. Preventing bird–window collisions. *Wilson J. Ornithol.* **2009**, *121*, 314–321. [CrossRef]
60. Brown, B.B.; Kusakabe, E.; Antonopoulos, A.; Siddoway, S.; Thompson, L. Winter bird-window collisions: Mitigation success, risk factors, and implementation challenges. *PeerJ* **2019**, *7*, e7620. [CrossRef]
61. Sheppard, C.D. Evaluating the relative effectiveness of patterns on glass as deterrents of bird collisions with glass. *Glob. Ecol. Conserv.* **2019**, *20*, e00795. [CrossRef]
62. Benedict, R.A.; Billeter, M.C. Discarded bottles as a cause of mortality in small vertebrates. *Southeast. Nat.* **2004**, *3*, 371–378. [CrossRef]
63. Roman, L.; Schuyler, Q.A.; Hardesty, B.D.; Townsend, K.A. Anthropogenic debris ingestion by avifauna in Eastern Australia. *PLoS ONE* **2016**, *11*, e0158343. [CrossRef] [PubMed]
64. Foley, C.J.; Feiner, Z.S.; Malinich, T.D.; Höök, T.O. A meta-analysis of the effects of exposure to microplastics on fish and aquatic invertebrates. *Sci. Total Environ.* **2018**, *631–632*, 550–559. [CrossRef] [PubMed]
65. Dawson, T.P.; Jackson, S.T.; House, J.I.; Prentice, I.C.; Mace, G.M. Beyond predictions: Biodiversity in a changing climate. *Science* **2011**, *332*, 53–58. [CrossRef]
66. Bellard, C.; Bertelsmeier, C.; Leadley, P.; Thuiller, W.; Courchamp, F. Impacts of climate change on the future of biodiversity. *Ecol. Lett.* **2012**, *15*, 365–377. [CrossRef]
67. Bernal, B.; Murray, L.T.; Pearson, T.R.H. Global carbon dioxide removal rates from forest landscape restoration activities. *Carbon Balance Manag.* **2018**, *13*, 22. [CrossRef] [PubMed]

Journal of
Zoological and Botanical Gardens

MDPI

Review

Southern Sea Otter Rehabilitation: Lessons and Impacts from the Monterey Bay Aquarium

Leilani Konrad [1,*], Jessica A. Fujii [1], Sandrine Hazan [1], Andrew B. Johnson [2], Karl A. Mayer [1], Michael J. Murray [1], Teri E. Nicholson [1], Michelle M. Staedler [1] and Colleen Young [3]

1 Monterey Bay Aquarium, 886 Cannery Row, Monterey, CA 93940, USA
2 Defenders of Wildlife, P.O. Box 401, Folsom, CA 95763, USA
3 Marine Wildlife Veterinary Care and Research Center, Office of Spill Prevention and Response, California Department of Fish and Wildlife, Santa Cruz, CA 95060, USA
* Correspondence: lkonrad@mbayaq.org

Abstract: As biodiversity continues to decline across the globe, conservation of wildlife species and the ecosystems they inhabit is more important than ever. When species dwindle, ecosystems that depend on them are also impacted, often leading to a decrease in the life-giving services healthy ecosystems provide to humans, wildlife, and the global environment. Methods of wildlife conservation are complex and multi-faceted, ranging from education and advocacy to, research, restoration, and rehabilitation. Here, we review a conservation program focused on helping recover the federally listed threatened southern sea otter (*Enhydra lutris nereis*) population. We describe the development of unique rehabilitation methods and steps taken to advance the program's conservation impact. Understanding this evolution can inform conservation efforts for other vulnerable species and their ecosystems.

Keywords: aquarium; collaboration; southern sea otter; surrogacy; wildlife rehabilitation

Citation: Konrad, L.; Fujii, J.A.; Hazan, S.; Johnson, A.B.; Mayer, K.A.; Murray, M.J.; Nicholson, T.E.; Staedler, M.M.; Young, C. Southern Sea Otter Rehabilitation: Lessons and Impacts from the Monterey Bay Aquarium. *J. Zool. Bot. Gard.* **2022**, *3*, 641–652. https://doi.org/10.3390/jzbg3040047

Academic Editors: Ursula Bechert and Debra C. Colodner

Received: 30 September 2022
Accepted: 17 November 2022
Published: 24 November 2022

1. Introduction

Conserving wildlife and ecosystems is increasingly important as biodiversity continues to decline across the planet. While the intrinsic value of wildlife is clear, key species are also vital in maintaining functioning ecosystems. One potential method to help restore diminished wildlife populations and their ecosystems is the rehabilitation and release of orphaned, ill, or injured animals [1]. Despite its highly complex and contextual nature, wildlife rehabilitation can serve as a conservation tool to support the recovery of threatened populations [1,2]. Understanding how the Monterey Bay Aquarium's Sea Otter Program has evolved to enhance conservation outcomes can inform future restoration efforts for other threatened species and their ecosystems.

1.1. Wildlife Rehabilitation

The treatment and care of injured, diseased, and displaced indigenous wildlife, and their subsequent release as healthy animals to native habitats in the wild [3] is a broadly used conservation strategy for marine and terrestrial species. Thousands of wildlife rehabilitation programs exist globally, and the practice continues to grow [4,5]. These programs include facilities and dedicated personnel to care for ill or injured wildlife, strategies to galvanize public interest in the welfare of local wildlife populations, opportunities to advance species-specific husbandry and veterinary care and methods to identify threats to wildlife populations [1].

While most of these rehabilitation programs have a common goal of supporting wildlife, their focus extends from the rescue and release of individual animals [6,7] to larger-scale reintroduction projects of threatened and endangered species [8,9]. This broad range in program focus is often influenced by available resources and the rehabilitation requirements of target animals. Because of their species-specific expertise, life-support systems and facilities, and financial resources, some zoos and aquariums are well suited to engage in wildlife rehabilitation. However, to achieve broader conservation goals, programs may need to establish partnerships to leverage external expertise and resources. These partnerships ultimately share or offset heavy fiscal burdens, inform strategies that minimize threats to individual animals or the wild population, and assist in identifying population-level conservation impacts.

By establishing these partnerships, a growing number of rehabilitation programs are demonstrating their effectiveness beyond individual animal welfare and measuring benefits to wild populations through post-release monitoring. For example, the California condor (*Gymnogyps californianus*) recovered from the brink of extinction through rescue, rehabilitation, and novel captive breeding and reintroduction techniques developed among zoos, non-profit organizations, and government agencies [10]. In Florida, the Manatee Rescue, Rehabilitation, and Release Program (MRP) is comprised of zoos and aquariums, agencies, academic institutions, and non-profit organizations. Through increased post-release monitoring and strategic interventions by MRP, 92% of subadult and adult West Indian manatees (*Trichechus manatus*) and a relatively high proportion of calves were rescued and released over a twenty-six-year period and successfully acclimated to the wild [11]. In the Hawaiian islands, approximately 30% of the endangered Hawaiian monk seal (*Neomanachus schauinsalndi*) population in 2012 was alive as the direct result of either opportunistic interventions or rescue and treatment by cooperating federal and state agencies and nonprofit partners [12]. These partnerships are examples of how rehabilitation programs can provide measurable benefits to wild populations.

1.2. The Southern Sea Otter

Following near-extirpation during the 18th and 19th century maritime fur trade, the southern sea otter (*Enhydra lutris nereis*) population has grown to 3000 individuals throughout central California [13]. Despite this increase, the current population only inhabits 13% of its historical range from Oregon to central Baja, Mexico (Figure 1) [14]. Since 1977, the southern sea otter remains listed as "threatened" under the Endangered Species Act (ESA), and the population's status and recovery plan are managed by the USFWS under the authority of the Marine Mammal Protection Act.

Sea otters are considered a keystone species, playing a significant role in restoring and maintaining the resilience of seagrass and kelp forest ecosystems through cascading trophic relationships with their invertebrate prey [15–20]. Because they reside at the interface between land and nearshore coastal waters, sea otters are susceptible to a wide variety of natural and anthropogenic threats, such as shark bites, parasites, toxins, and infectious diseases that may lead to stranding [21,22]. During their first six months, sea otter pups depend on their mothers to nurse and nourish them, and teach them survival skills, such as foraging, grooming, socializing, and avoiding threats [23,24]. If a sea otter mother cannot find enough food to meet the high energetic requirements of pup rearing [23] or becomes ill or injured, she may abandon her pup prematurely [25,26], threatening its survival without human intervention.

Determining trends in sea otter strandings (e.g., cause and demography) is critical to identifying threats to wild populations. The California Department of Fish and Wildlife (CDFW; formerly CA Department of Fish and Game) began responding to and systematically documenting southern sea otter strandings in 1968. Since then, a network of collaborators including CDFW, United States Geological Survey (USGS), Monterey Bay Aquarium (MBA), and the Marine Mammal Center (TMMC) have worked together to collect and examine stranded sea otters. Most sea otter carcasses receive a necropsy by CDFW pathologists who, if possible, determine a primary cause of death. Sea otters that strand alive are generally collected by MBA, TMMC, or CDFW and evaluated at MBA or TMMC. TMMC and MBA are the only permitted facilities that currently rehabilitate southern sea otters in California. Therefore any sea otters who are candidates for rehabilitation are reared and cared for by TMMC and MBA until they are released back to the wild or deemed non-releasable and transferred to a long-term care facility.

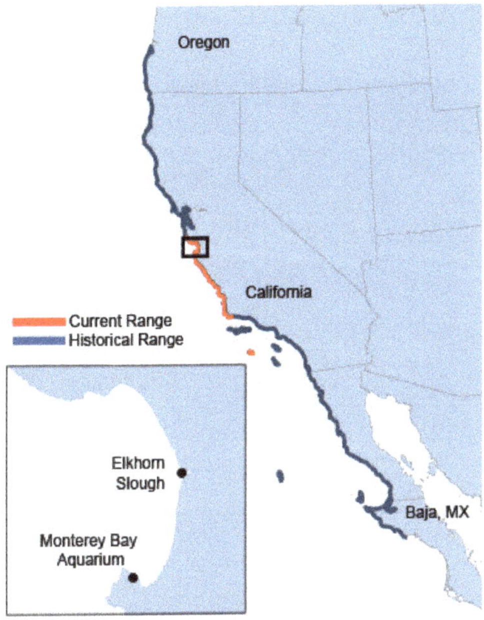

Figure 1. The current and historical range extent for the southern sea otter subspecies and (inset) location of Elkhorn Slough in relation to the Monterey Bay Aquarium.

For nearly four decades, the Monterey Bay Aquarium's Sea Otter Program has rescued, treated, and released stranded southern sea otters [16]. Reviewing this program's evolution highlights how wildlife rehabilitation potentially can support not only species recovery, but also ecosystem restoration more broadly and affirms how partnerships and collaborations may be leveraged for success.

2. Materials and Methods

2.1. The Monterey Bay Aquarium

The Monterey Bay Aquarium's mission is to inspire conservation of the ocean. Since opening its doors in 1984, MBA has advanced conservation through a fleet of animal husbandry, communications, education, exhibition, guest experience, marketing, policy, and research programs aimed at restoring and protecting California's ocean and coastal ecosystems. MBA has accomplished some of this work through rehabilitation efforts that advance the conservation of three imperiled California wildlife populations: the green sea turtle (*Chelonia mydas*), the Western snowy plover (*Charadrius nivosus nivosus*), and the southern sea otter. Although MBA engages in a variety of activities to support its conservation goals, in this paper we focus on its contributions to southern sea otter recovery through its Sea Otter Program.

2.2. Methods

The MBA Sea Otter Program (hereafter, 'program') consists of multiple components including stranding response, rehabilitation, population monitoring, research, husbandry, and veterinary care. To achieve the program's overall goal of southern sea otter recovery during the last 38 years, these activities have been conducted in collaboration with state and federal agencies, other non-profits, and universities. Although all of the program's efforts are equally important, this review focuses on the history of sea otter rehabilitation at

MBA with an emphasis on aspects that may inform other wildlife rehabilitation programs. We do so by highlighting successes and failures at critical stages that were formative in the program's evolution (Figure 2), summarizing southern sea otter stranding data, and providing several examples of how collaboration has informed advancement in scientific research.

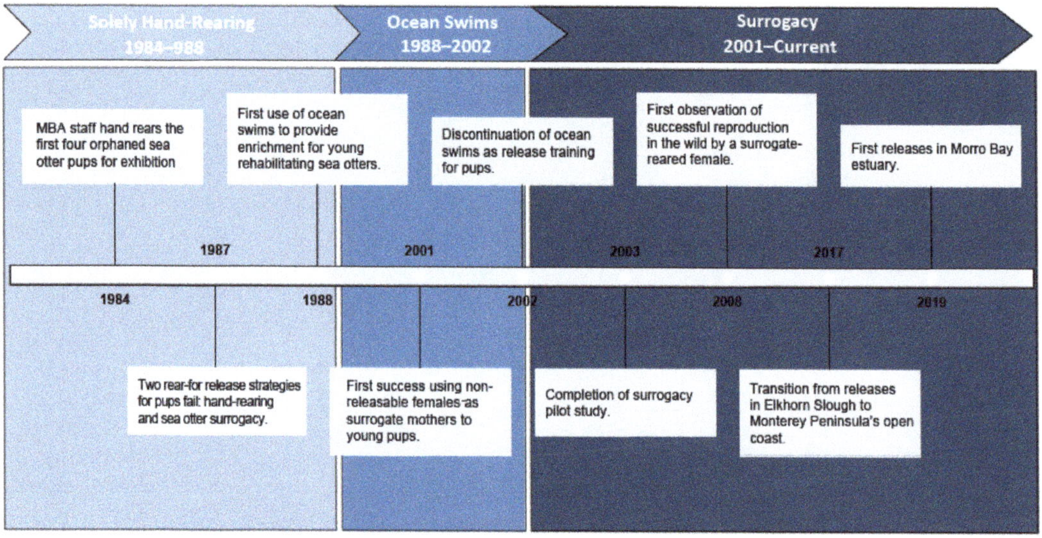

Figure 2. A timeline of key successes and failures through the history of MBA's Sea Otter Program rehabilitation efforts, marked by transitions in methods (i.e., hand-rearing, ocean swims, surrogacy).

3. Results

3.1. The Program's Beginnings

MBA rescued its first sea otters, when four pups separately stranded along the central California coast between February and April of 1984, half a year before the aquarium opened to the public (Figure 2). Aquarium staff hand-reared these pups, and because of this intensive care by humans, the USFWS deemed the pups non-releasable and authorized MBA to provide long-term care while exhibiting the otters. After acquiring these initial animals, the program continued to rescue, care for, and when possible, release stranded sea otters of all ages. While it was not the original plan to rehabilitate or exhibit sea otters in this new aquarium, support for the local sea otter population and the public's desire to help stranded individuals was clear.

As the number of sea otters in need of rescue continued to increase (Figure 3), program strategies had to evolve. Although animals of all age classes stranded, the majority of rescued older sea otters were in very poor health condition and only 22% (n = 146) were able to be rehabilitated from a wide variety of stranding causes (Table 1).

By contrast, pups stranded in relatively good health and responded well to treatment. With limited options for long-term placement of orphaned pups at other zoos and aquariums, and a wild population still well below recovery, program staff had to explore strategies for rearing pups for release to the wild. At the time, sea otter rearing strategies were unknown, but staff explored two different rear-for-release methods in 1987 (Figure 2). One case involved hand-rearing a pup prior to release, and in the other, an attempt was made to establish a bond between a pup and a resident non-releasable adult female sea otter. Neither strategy was successful, and program staff decided to continue to provide human care but hypothesized that exposing pups to their natural environment during development could aid in achieving more successful release outcomes. To do so, staff established strong bonds with individual animals and conducted frequent ocean enrichment swims, which allowed the pups to explore nearshore rocky reefs and subtidal kelp forests in Monterey Bay. The intent of these swims was to encourage prey identification, foraging, and socialization with other wild otters in preparation for release as juveniles. Although ocean swims provided enrichment and allowed young animals to develop diving, swimming, and foraging skills, the close connection with human caregivers prevented many of the pups from establishing a natural wariness of people and reintegrating with the wild population after release [25].

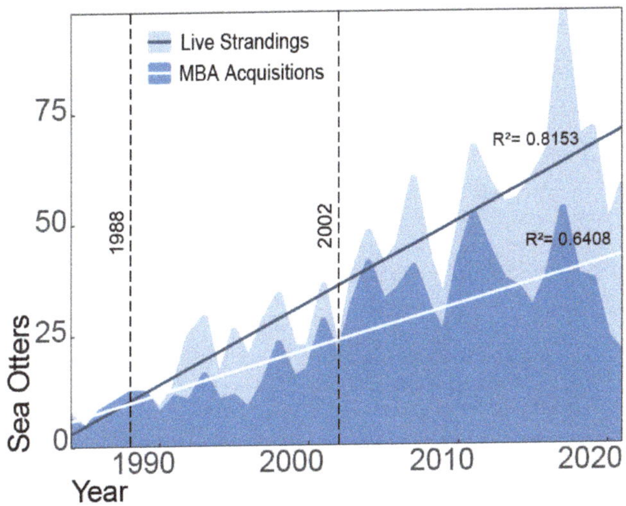

Figure 3. The number of sea otters that strand alive has increased over time, with annual fluctuations (R^2 = 0.8153). Likewise, the number of strandings responded to by the program has also increased over it's history (R^2 = 0.6408). From the start of the program, the caseload has increased from five animals in 1984 to nineteen at the start of surrogacy methods in 2002. From 1988 to 2002, when the program focused on hand-rearing and conducting enrichment ocean swims with sea otter pups, staff managed a caseload total of 213 animals, averaging 15 sea otters a year.

Table 1. Stranding causes of adult, subadult, and juvenile sea otters rescued, rehabilitated, and released throughout the program's history (*n* = 146, see also Nicholson et al. 2018).

Stranding Cause	Description	No. Cases
Pre-mature weaning	Signs of emaciation or stunting due to early weaning or chronic malnutrition during weaning	38
ELS/mating trauma	Significant abrading or loss of nose-pad due to mating trauma (28), emaciation related to late-lactation (3)	31
Anthropogenic Trauma	Laceration, punctures, and fractures consistent with boat strike (2), fisheries interactions/line entanglement (6), tar or oiling (2), nonspecific anthropogenic trauma (5)	15
Neurological disease	Fine muscle tremors, seizures, and/or loss of motor function	14
Acanthocephalan peritonitis	Presence of acanthocephalan parasites	11
Shark bite	Lacerations from attempted depredation	11
Geriatric	Pathology related to old age and poor dentition	4
Other	Storm stress (1), respiratory disease (2), reproductive complications (1) or unknown (18)	22
Total		146

To closely monitor outcomes in the wild, sea otters receive intraperitoneal radio transmitter implants [27], and staff track them intensively over a two-week post-release period. Overall, sea otters released to the wild survived at rates comparable to their wild counterparts [25], except for young pups. By comparison, only 27% of individuals stranding as pups successfully reacclimated to the wild, while most required recapture and permanent placement in zoological facilities because of a failure to forage successfully or maintain a wariness of people [25,28].

As program staff explored how to successfully return young otters to the wild, its caseload of all ages continued to increase from six animals in 1984 to nineteen in 2002 (Figure 3). Despite the labor-intensive nature of veterinary treatment, care, and enrichment swims, the program admitted 213 live-stranded sea otters during this period (1988–2002), rehabilitating and releasing just under 50%. Through ongoing studies of the wild population, post-release tracking, and accumulation of data quantifying survival and reproduction rates of released individuals of all ages, program staff identified the need for a shift in rehabilitation methods to increase successful outcomes of stranded pups. Additionally, this decision process was aided by findings from a Blue Ribbon Panel [29] (a panel of subject matter experts) in 2002, which provided a framework for caseload management and clear integration of research and conservation priorities to best utilize limited resources.

3.2. *Surrogacy*

After an early failure while pairing a pup with an adult female sea otter, starting in 2001 the program re-explored using non-releasable females as surrogate mothers to rehabilitate dependent pups. This strategy was reimplemented to limit interactions with humans and leverage the natural maternal instincts of adult females to provide species-specific care to sea otter pups. Based on historical cases of sea otter adoption [30,31], staff believed this method would allow pups to better integrate with the wild population and avoid humans following release. The first successful introduction between an orphaned pup and a wild female in long-term care at MBA occurred in 2001 (Figure 2). Since then, the program has used surrogacy as a method to rear stranded pups for release.

Sea otter surrogacy involves five key stages; (1) stranding response, (2) stabilization, (3) surrogate rearing, (4) release preparation, and (5) post-release monitoring [32]. During stabilization, staff provide hands-on care while wearing disguises to minimize pup habituation with humans. At this stage, pups develop basic grooming, diving, and foraging skills before introductions to a surrogate female sea otter at approximately 8–10 weeks of age [16]. Once introductions are successful, mother and pup remain together during dependency with limited human intervention [16]. At around 6 months of age, the pup is weaned (i.e., permanently separated) from its surrogate, and veterinary staff administer several health exams in preparation for release [16]. Along with a VHF radio transmitter, released otters are instrumented with a unique color and placement combination of hind-flipper tags for identification in the field [32,33]. To assess how individuals are adjusting to the wild, post-release monitoring in collaboration with TMMC, CDFW, and USGS, details an otter's daily location, distance traveled, foraging success, behavior, and body condition.

With the development of these protocols, the focus of the program's pup-rearing efforts shifted from human-facilitated enrichment in the natural environment to surrogate-fostering of natural behaviors such as diving, grooming, foraging, and socializing. To assess whether this new method would result in greater success for released individuals compared with previous hand-rearing methods, the program conducted a trial study of five surrogate-reared pups that stranded from 2001 to 2003 [25] and were subsequently released into Elkhorn Slough. This release site is a seven-mile-long estuary located approximately 20 miles north of MBA along the center of the Monterey Bay coastline as shown in Figure 1.

During the study, surrogate-reared pups integrated well into the wild population, foraged successfully, and maintained their wariness of people [25]. The overall survival rate one-year post release was 71%, which was significantly higher than the rate of success (27%) prior to surrogacy [25]. These findings supported the shift in the program's approach to rearing stranded pups and demonstrated the importance of post-release monitoring and the scientific method for informing program development. Over the course of the program's history, MBA has responded to 80% of all live stranded pups (Figure 4), and in recent years, as the numbers of live stranded sea otters of all age classes has increased, collaborators such as TMMC have been rehabilitating more mature animals, allowing MBA to focus on increasing its capacity for rehabilitating pups.

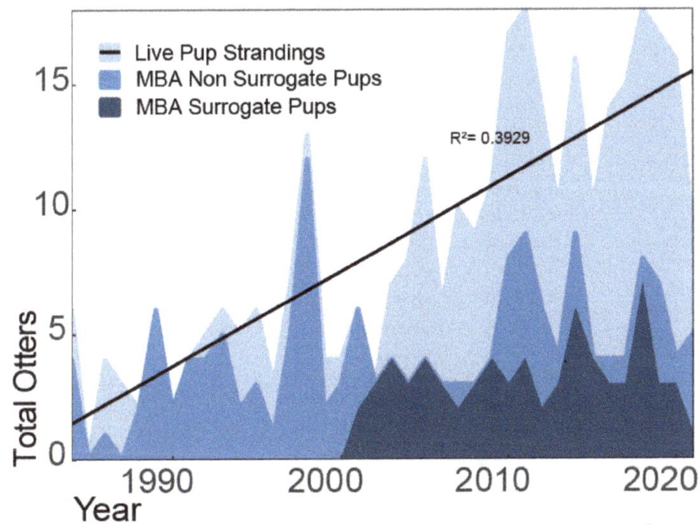

Figure 4. Throughout the program's history, live pup strandings have steadily increased (R^2 = 0.3929), and MBA responded to approximately 80% of these events. 50% of orphaned pups received extended care at the aquarium prior to release or transfer to a long-term care facility. From its implementation in 2002, only 28% of qualified pups (aged <10 weeks at stranding) were surrogate-reared. Criteria for aging pups at this young stage included size (total length), weight, tooth eruption patterns, and presence of natal pelage. Based on these methods, pup aging error is estimated at +/− 2 weeks.

3.3. Conservation Impact of Releasing Surrogate-Reared Sea Otters

Through rigorous data collection, a series of collaborative scientific research projects documented the positive conservation impacts of this program on local populations and ecosystems. By conducting long-term monitoring of released sea otters (n = 37) from 2002 to 2015 as well as annual monitoring of the population of sea otters in Elkhorn Slough, researchers were able to model the effect of this rehabilitation program on population growth [16]. In 2002, the population in Elkhorn Slough was approximately 20 individuals, and by 2015 it had grown to nearly 150 [34]. Based on release numbers and demographics, as well as observed and estimated survival and reproductive rates of rehabilitated and wild sea otters, surrogate-reared animals and their offspring were estimated to account for 55% of overall population growth in Elkhorn Slough [16]. Additionally, the release of female sea otters to this area likely contributed to a shift in the population from an area dominated by transient males, to a stable reproductive population. In much the same way that the presence of sea otters helps maintain kelp forests, increasing numbers of sea otters in Elkhorn Slough supported eelgrass (*Zostera* spp.) growth and recovery from eutrophic conditions, aiding in ecosystem restoration for other species in this habitat [17]. Specifically researchers identified a trophic cascade involving sea otters, crabs, and sea slugs that enhanced grazing of epiphytic algae from eelgrass blades, improving health and density of seagrass meadows [17]. Compared to measurements prior to the recolonization of sea otters, eelgrass extent increased 60% [35]. These outcomes have provided scientific support for surrogacy as a rehabilitation method and for releasing these rehabilitated individuals to enhance population recovery and ecosystem restoration.

3.4. Collaborative Conservation Research

Beyond the direct population and ecosystem benefits of releasing sea otters, rehabilitation has also contributed to conservation efforts by informing research and development of veterinary and animal welfare protocols. This includes clinical diagnosis and treatment [36–39], adaptation of methods from techniques and technologies of other aspects of medicine to sea otters [40–44], expansion of the idiosyncrasies of pharmacology [45], and development/maintenance of a robust archive of biological materials sampled from sea otters over several decades [46–49].

Rehabilitation of sea otters at MBA has also inspired research to advance sea otter husbandry [50], oiled animal care [39,51], and knowledge of sea otter biology and physiology [23,41,52–55]. Findings from these collaborative research projects may inform wildlife managers about how best to anticipate, overcome, and prevent, risks to sea otters throughout a range of scenarios within wild and long-term care settings.

Sea otters undergoing rehabilitation have participated in collaborative experimental research that would be impossible to conduct in the wild. For example, during controlled exposure within monitored pools, sea otters demonstrated their vulnerability to drowning within commercial finfish and shellfish traps, which was a suspected population threat. This result led to a successful policy change that required the modification of fishing equipment with the addition of a rigid 5-inch ring to fyke openings, excluding most otters from entering and likely reducing sea otter entrapment [56]. These research advancements have either informed or directly aided sea otter conservation efforts and show how collaboration can magnify the benefits of wildlife rehabilitation.

4. Conclusions

4.1. Key Takeaway

MBA's Sea Otter Program exemplifies lessons learned and knowledge attained through rigorous research, methodological adaptation, and collaboration. Understanding how these factors have shaped the evolution of MBA's program can be helpful to other rehabilitation and conservation partnerships with similar goals [28]. The dissemination of information among programs can aid in addressing challenges in wildlife rehabilitation. The resource-intensive nature of wildlife rehabilitation is always a major obstacle, but collaboration provides alternative solutions and extends resources to achieve program goals. For MBA and other programs, increasing collaboration and partnerships continue to advance methods of rehabilitation, post-release monitoring, and analysis of impacts to wild populations, informing and improving conservation efforts into the future.

4.2. The Future

The evolution of this program resulted from a need to address the increase in live sea otter strandings, as well as the growing knowledge that rehabilitation methods should be informed by a scientific approach focused on improving individual animal and population welfare, and carried out in partnership with other groups and organizations. As the program continues to evolve, this knowledge and approach will be important when addressing future challenges in sea otter conservation.

Since the study at Elkhorn Slough, MBA's rehabilitation program and its partners have continued to successfully release animals in estuaries and open coast kelp forests within the southern sea otter's current range. Long-term survival of and reproduction by released sea otters have contributed to their species' keystone role of promoting healthier seagrass and kelp forest ecosystems [17,18,20]. These ecosystem benefits indicate that expanding releases for surrogate-reared sea otters could have local population-level and ecosystem benefits in other areas throughout the California coastline. Continued post-release monitoring of wild sea otters is essential to understand how releasing otters within historical areas could aid population recovery and nearshore ecosystem restoration.

Although sea otter population abundance has increased along the central coast of California, the range extent of the population has remained mostly unchanged for the last decade (Figure 5). The growth of the existing wild population is limited by the lack of range expansion likely caused by shark bite mortality at the northern and southern peripheries [13]. Because range expansion may be critical for achieving population growth, MBA and its partners must focus their rehabilitation and release efforts in areas where population numbers are low to contribute to southern sea otter recovery.

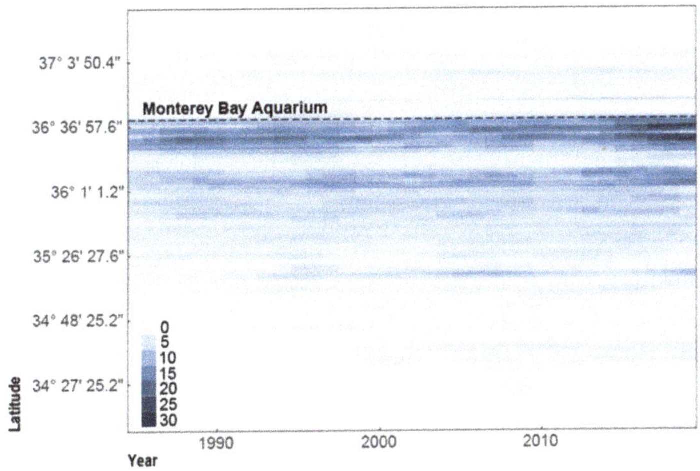

Figure 5. Southern sea otter population density; measured as otters per square kilometer across the extent of the subspecies range from 1985–2019. MBA is located at 36°37′4.8″ latitude [34].

Adult sea otters exhibit strong site fidelity [57], which creates challenges when attempting to rehabilitate and release them outside of the current range. MBA's method of releasing surrogate-reared juveniles could be a useful approach in addressing the population's lack of range expansion. Because an estimated 42% of releasable live strandings are young pups without an established home range, this age class presents an impactful rehabilitation opportunity. Since the start of the surrogacy program in 2002, MBA has only reared 28% of live stranded pups for release (Figure 4). Further partnerships with other zoos and aquariums could increase the capacity for surrogate-rearing orphaned pups, which may aid southern sea otter range expansion efforts. As the program continues to develop, existing and emerging partnerships will be vital in understanding if surrogacy could be applied to sea otter reintroduction efforts to advance sea otter recovery under the ESA.

Author Contributions: Conceptualization, J.A.F., L.K. and T.E.N.; Methodology, J.A.F., L.K. and C.Y.; Data Curation, L.K., T.E.N. and C.Y.; Writing—Original Draft Preparation, J.A.F., L.K. and T.E.N.; Writing—Review & Editing, J.A.F., S.H., A.B.J., L.K., K.A.M., M.J.M., T.E.N., M.M.S. and C.Y.; Visualization, L.K.; Supervision, J.A.F. and T.E.N. All authors have read and agreed to the published version of the manuscript.

Funding: This research was supported by generous donor contributions to the Monterey Bay Aquarium.

Institutional Review Board Statement: This study operated under U.S. Fish & Wildlife Service permits #MA032027-2 and LOA032027-2, and Monterey Bay Aquarium permits #MA186914 and #032027. All animal handling was in accordance with the requirements of USDA Class R Research Facility license # 93-R-0476.

Data Availability Statement: Not applicable.

Acknowledgments: Jack Ames, Greta Austin, Gena Bentall, Allie Bondi-Taylor, Sue Campbell, Dave Casper, Jennifer Coffey, Chris DeAngelo, James Estes, Katie Finch, Krista Hanni, Mike Harris, Katie Hawkins, Julie Hymer, Michele Jefferies, Sonny Knaub, Erin Lenihan, Kaitlyn Reilly, Marianne Riedman, Julie Stewart, Kattie Stong, Candace Tahara, Marcie Tarvid, Bob VanWagenen, Thomas Williams, Linda Yingling, Marissa Young, and the many staff, volunteers, and interns past and present at MBA and TMMC who helped care for, and track the many animals in the program.

Conflicts of Interest: The authors declare no conflict of interest.

References

1. Paterson, J.E.; Carstairs, S.; Davy, C.M. Population-level effects of wildlife rehabilitation and release vary with life-history strategy. *J. Nat. Conserv.* **2021**, *61*, 125983. [CrossRef]
2. Saran, K.A.; Parker, G.; Parker, R.; Dickman, C.R. Rehabilitation as a conservation tool: A case studyusing the common wombat. *Pac. Conserv. Biol.* **2011**, *17*, 310–319. [CrossRef]
3. Miller, E. *Minimum Standards for Wildlife Rehabilitation*, 4th ed.; National Wildlife Rehabilitators Association: St. Cloud, MN, USA, 2012.
4. Kelly, G. Importance of taxonomic group, life stage and circumstance of rescue upon wildlife rehabilitation in Ontario, Canada. *J. Nat. Conserv.* **2020**, *57*, 125897. [CrossRef]
5. Molony, S.; Baker, P.; Garland, L.; Cuthill, I.; Harris, S. Factors that can be used to predict release rates for wildlife casualties. *Anim. Welf.* **2007**, *16*, 361.
6. Guy, A.J.; Curnoe, D.; Banks, P.B. A survey of current mammal rehabilitation and release practices. *Biodivers. Conserv.* **2013**, *22*, 825–837. [CrossRef]
7. Karesh, W.B. Wildlife rehabilitation: Additional considerations for developing countries. *J. Zoo Wildl. Med.* **1995**, *26*, 2–9.
8. Astore, V.; Estrada, R.; Jacome, N.L. Reintroduction strategy for the Andean condor conservation program, Argentina. *Int. Zoo. Yearb.* **2017**, *51*, 124–136. [CrossRef]
9. Hogg, C.; Lee, A.; Srb, C.; Hibbard, C. Metapopulation management of an endangered species with limited genetic diversity in the presence of disease: The Tasmanian devil Sarcophilus harrisii. *Int. Zoo Yearb.* **2017**, *51*, 137–153. [CrossRef]
10. Walters, J.R.; Derrickson, S.R.; Michael Fry, D.; Haig, S.M.; Marzluff, J.M.; Wunderle, J.M. Status of the California condor (*Gymnogyps californianus*) and Efforts to Achieve Its Recovery. *Auk* **2010**, *127*, 969–1001. [CrossRef]
11. Adimey, N.M.; Ross, M.; Hall, M.; Reid, J.P.; Barlas, M.E.; Diagne, L.W.K.; Bonde, R.K. Twenty-six years of post-release monitoring of Florida manatees (*Trichechus manatus latirostris*): Evaluation of a cooperative rehabilitation program. *Aquat. Mamm.* **2016**, *42*, 391. [CrossRef]
12. Harting, A.L.; Johanos, T.C.; Littnan, C.L. Benefits derived from opportunistic survival-enhancing interventions for the Hawaiian monk seal: The silver BB paradigm. *Endanger. Species Res.* **2014**, *25*, 89–96. [CrossRef]
13. Hatfield, B.B.; Yee, J.L.; Kenner, M.C.; Tomoleoni, J.A. *California Sea Otter (Enhydra lutris nereis) Census Results, Spring 2019*; U.S. Geological Survey Data Series; U.S. Geological Survey National Center: Reston, VA, USA, 2019.
14. United States Fish and Wildlife Service. *Southern Sea Otter: 5-Year Review: Summary and Evaluation*; U.S. Fish and Wildlife Service Ventura Fish and Wildlife Office: Ventura, CA, USA, 2015; Volume 42.
15. Estes, J.A.; Palmisano, J.F. Sea otters: Their role in structuring nearshore communities. *Science* **1974**, *185*, 1058–1060. [CrossRef]
16. Mayer, K.A.; Tinker, M.T.; Nicholson, T.E.; Murray, M.J.; Johnson, A.B.; Staedler, M.M.; Fujii, J.A.; Van Houtan, K.S. Surrogate rearing a keystone species to enhance population and ecosystem restoration. *Oryx* **2021**, *55*, 535–545. [CrossRef]
17. Hughes, B.B.; Eby, R.; Van Dyke, E.; Tinker, M.T.; Marks, C.I.; Johnson, K.S.; Wasson, K. Recovery of a top predator mediates negative eutrophic effects on seagrass. *PNAS* **2013**, *110*, 15313–15318. [CrossRef]
18. Smith, J.G.; Tomoleoni, J.; Staedler, M.; Lyon, S.; Fujii, J.; Tinker, M.T. Behavioral responses across a mosaic of ecosystem states restructure a sea otter–urchin trophic cascade. *PNAS* **2021**, *118*, e2012493118. [CrossRef]
19. Kenner, M.C.; Tinker, M.T. Stability and change in kelp forest habitats at San Nicolas Island. *West. N. Am. Nat.* **2018**, *78*, 633–643. [CrossRef]
20. Estes, J.A.; Terborgh, J.; Brashares, J.S.; Power, M.E.; Berger, J.; Bond, W.J.; Carpenter, S.R.; Essington, T.E.; Holt, R.D.; Jackson, J.B. Trophic downgrading of planet Earth. *Science* **2011**, *333*, 301–306. [CrossRef]
21. Miller, M.A.; Moriarty, M.E.; Henkel, L.; Tinker, M.T.; Burgess, T.L.; Batac, F.I.; Dodd, E.; Young, C.; Harris, M.D.; Jessup, D.A. Predators, disease, and environmental change in the nearshore ecosystem: Mortality in southern sea otters (*Enhydra lutris nereis*) from 1998–2012. *Front. Mar. Sci.* **2020**, *7*, 582. [CrossRef]
22. Nicholson, T.E.; Mayer, K.A.; Staedler, M.M.; Fujii, J.A.; Murray, M.J.; Johnson, A.B.; Tinker, M.T.; Van Houtan, K.S. Gaps in kelp cover may threaten the recovery of California sea otters. *Ecography* **2018**, *41*, 1751–1762. [CrossRef]
23. Thometz, N.; Staedler, M.; Tomoleoni, J.; Bodkin, J.L.; Bentall, G.; Tinker, M.T. Trade-offs between energy maximization and parental care in a central place forager, the sea otter. *Behav. Ecol.* **2016**, *27*, 1552–1566. [CrossRef]
24. Cortez, M.; Wolt, R.; Gelwick, F.; Osterrieder, S.K.; Davis, R.W. Development of an altricial mammal at sea: I. Activity budgets of female sea otters and their pups in Simpson Bay, Alaska. *J. Exp. Mar. Biol. Ecol.* **2016**, *481*, 71–80. [CrossRef]

25. Nicholson, T.E.; Mayer, K.A.; Staedler, M.M.; Johnson, A.B. Effects of rearing methods on survival of released free-ranging juvenile southern sea otters. *Biol. Conserv.* **2007**, *138*, 313–320. [CrossRef]
26. Riedman, M.L.; Estes, J.A.; Staedler, M.M.; Giles, A.A.; Carlson, D.R. Breeding patterns and reproductive success of California sea otters. *J. Wildl. Manag.* **1994**, *58*, 391–399. [CrossRef]
27. Ralls, K.; Siniff, D.B.; Williams, T.D.; Kuechle, V. An intraperitoneal radio transmitter for sea otters. *Mar. Mamm. Sci.* **1989**, *5*, 376–381. [CrossRef]
28. Johnson, A.; Mayer, K. The value of rescuing, treating, and releasing live-stranded sea otters. In *Sea Otter Conservation*; Elsevier: Amsterdam, The Netherlands, 2015; pp. 235–255.
29. Tama, J. Crises, commissions, and reform: The impact of blue-ribbon panels. *PRQ* **2014**, *67*, 152–164. [CrossRef]
30. Kenyon, K. *The Sea Otter in the Eastern North Pacific*; Dover Publications: New York, NY, USA, 1975.
31. Staedler, M.M.; Riedman, M.L. A case of adoption in the California Sea Otter. *Mar. Mamm. Sci.* **1989**, *5*, 391–394. [CrossRef]
32. Nicholson, T.; Mayer, K.; Hazan, S.; Murray, M.; Van Houtan, K.S.; DeAngelo, C.; Fujii, J. Advancing surrogate-rearing methods to enhance southern sea otter recovery. *Biol. Conserv.* 2022; *in review*.
33. Siniff, D.B.; Ralls, K. Reproduction, survival and tag loss in California sea otters. *Mar. Mamm. Sci.* **1991**, *7*, 211–299. [CrossRef]
34. Tinker, M.T.; Hatfield, B.B. *Annual California Sea Otter Census—1985–2019 Spring Census Summary*; U.S. Geological Survey: Santa Cruz, CA, USA, 2019. [CrossRef]
35. Hughes, B.B.; Wasson, K.; Tinker, M.T.; Williams, S.L.; Carswell, L.P.; Boyer, K.E.; Beck, M.W.; Eby, R.; Scoles, R.; Staedler, M. Species recovery and recolonization of past habitats: Lessons for science and conservation from sea otters in estuaries. *PeerJ* **2019**, *7*, e8100. [CrossRef]
36. Bowen, L.; Miles, A.K.; Murray, M.; Haulena, M.; Tuttle, J.; Van Bonn, W.; Adams, L.; Bodkin, J.L.; Ballachey, B.; Estes, J. Gene transcription in sea otters (*Enhydra lutris*); development of a diagnostic tool for sea otter and ecosystem health. *Mol. Ecol. Resour.* **2012**, *12*, 67–74. [CrossRef]
37. Hanni, K.D.; Mazet, J.A.; Gulland, F.M.; Estes, J.; Staedler, M.; Murray, M.J.; Miller, M.; Jessup, D.A. Clinical pathology and assessment of pathogen exposure in southern and Alaskan sea otters. *J. Wildl. Dis.* **2003**, *39*, 837–850. [CrossRef]
38. Tyrrell, L.P.; Newsome, S.D.; Fogel, M.L.; Viens, M.; Bowden, R.; Murray, M.J. Vibrissae growth rates and trophic discrimination factors in captive southern sea otters (*Enhydra lutris nereis*). *J. Mammal.* **2013**, *94*, 331–338. [CrossRef]
39. Jessup, D.A.; Yeates, L.C.; Toy-Choutka, S.; Casper, D.; Murray, M.J.; Ziccardi, M.H. Washing oiled sea otters. *Wildl. Soc. Bull.* **2012**, *36*, 6–15. [CrossRef]
40. Murray, M.; Young, M.; Santymire, R. Use of the ACTH challenge test to identify the predominant glucocorticoid in the southern sea otter (*Enhydra lutris nereis*). *Conserv. Physiol.* **2020**, *8*, coz116. [CrossRef] [PubMed]
41. Moriarty, M.E.; Tinker, M.T.; Miller, M.A.; Tomoleoni, J.A.; Staedler, M.M.; Fujii, J.A.; Batac, F.I.; Dodd, E.M.; Kudela, R.M.; Zubkousky-White, V. Exposure to domoic acid is an ecological driver of cardiac disease in southern sea otters☆. *Harmful Algae* **2021**, *101*, 101973. [CrossRef] [PubMed]
42. Gunther-Harrington, C.T.; Moriarty, M.E.; Field, C.L.; Adams, L.M.; Johnson, C.K.; Murray, M.J. Transthoracic echocardiographic evaluation and serum cardiac troponin values in anesthetized healthy female southern sea otters (*Enhydra lutris nereis*). *JZWM* **2021**, *52*, 490–498. [CrossRef]
43. Nicholson, T.E.; Mayer, K.A.; Staedler, M.M.; Gagné, T.O.; Murray, M.J.; Young, M.A.; Tomoleoni, J.A.; Tinker, M.T.; Van Houtan, K.S. Robust age estimation of southern sea otters from multiple morphometrics. *Ecol. Evol.* **2020**, *10*, 8592–8609. [CrossRef]
44. Bodkin, J.; Ballachey, B.; Esslinger, G. Synthesis of nearshore recovery following the 1989 Exxon Valdez oil spill: Trends in sea otter population abundance in Western Prince William Sound. *Exxon Valdez* **2011**, *1-17*, 1–17.
45. Lee, E.; Byrne, B.A.; Young, M.; Murray, M.; Miller, M.; Tell, L.A. Pharmacokinetic indices for cefovecin after single-dose administration to adult sea otters (*Enhydra lutris*). *JVPT* **2016**, *39*, 625–628. [CrossRef]
46. Miller, M.A.; Moriarty, M.E.; Duignan, P.J.; Zabka, T.S.; Dodd, E.; Batac, F.I.; Young, C.; Reed, A.; Harris, M.D.; Greenwald, K. Clinical signs and pathology associated with domoic acid toxicosis in southern sea otters (*Enhydra lutris nereis*). *Front. Mar. Sci.* **2021**, *8*, 585501. [CrossRef]
47. Shapiro, K.; Miller, M.A.; Packham, A.E.; Aguilar, B.; Conrad, P.A.; Vanwormer, E.; Murray, M.J. Dual congenital transmission of Toxoplasma gondii and Sarcocystis neurona in a late-term aborted pup from a chronically infected southern sea otter (*Enhydra lutris nereis*). *Parasitology* **2016**, *143*, 276–288. [CrossRef] [PubMed]
48. Bowen, L.; Miles, A.K.; Kolden, C.A.; Saarinen, J.A.; Bodkin, J.L.; Murray, M.J.; Tinker, M.T. Effects of wildfire on sea otter (*Enhydra lutris*) gene transcript profiles. *Mar. Mamm. Sci.* **2015**, *31*, 191–210. [CrossRef]
49. Miles, A.K.; Bowen, L.; Ballachey, B.; Bodkin, J.L.; Murray, M.; Estes, J.; Keister, R.A.; Stott, J.L. Variations of transcript profiles between sea otters *Enhydra lutris* from Prince William Sound, Alaska, and clinically normal reference otters. *Mar. Ecol. Prog.* **2012**, *451*, 201–212. [CrossRef]
50. Casson, C.J.; Murray, M.; Johnson, A.; Belting, T. *Sea Otter (Enhydra lutris) Care Manual*; Association of Zoos and Aquariums: Silver Spring, MD, USA, 2019.
51. Oiled Wildlife Care Network. *Protocols for the Care of Oil-Affected Sea Otters (Enhydra lutris)*; Oiled Wildlife Care Network: Davis, CA, USA, 2020.

52. Strobel, S.M.; Miller, M.A.; Murray, M.J.; Reichmuth, C. Anatomy of the sense of touch in sea otters: Cutaneous mechanoreceptors and structural features of glabrous skin. *Anat. Rec.* **2022**, *305*, 535–555. [CrossRef] [PubMed]
53. Beichman, A.C.; Koepfli, K.-P.; Li, G.; Murphy, W.; Dobrynin, P.; Kliver, S.; Tinker, M.T.; Murray, M.J.; Johnson, J.; Lindblad-Toh, K. Aquatic adaptation and depleted diversity: A deep dive into the genomes of the sea otter and giant otter. *Mol. Biol. Evol.* **2019**, *36*, 2631–2655. [CrossRef]
54. Wright, T.; Davis, R.W.; Pearson, H.C.; Murray, M.; Sheffield-Moore, M. Skeletal muscle thermogenesis enables aquatic life in the smallest marine mammal. *Science* **2021**, *373*, 223–225. [CrossRef]
55. Strobel, S.M.; Moore, B.A.; Freeman, K.S.; Murray, M.J.; Reichmuth, C. Adaptations for amphibious vision in sea otters (Enhydra lutris): Structural and functional observations. *J. Comp. Physiol. A* **2020**, *206*, 767–782. [CrossRef]
56. Hatfield, B.B.; Ames, J.A.; Estes, J.A.; Tinker, M.T.; Johnson, A.B.; Staedler, M.M.; Harris, M.D. Sea otter mortality in fish and shellfish traps: Estimating potential impacts and exploring possible solutions. *Endanger. Species. Res.* **2011**, *13*, 219–229. [CrossRef]
57. Ralls, K.; Eagle, T.C.; Siniff, D.B. Movement and spatial use patterns of California sea otters. *Can. J. Zool.* **1996**, *74*, 1841–1849. [CrossRef]

Journal of *Zoological and Botanical Gardens*

Review

Ex Situ Breeding Program with Wild-Caught Founders Provides the Source for Collaborative Effort to Augment Threatened New England Cottontail Populations

Jessie Brown * and Lauren Puccia *

Roger Williams Park Zoo, Providence, RI 02907, USA
* Correspondence: jbrown@rwpzoo.org (J.B.); lpuccia@rwpzoo.org (L.P.); Tel.: +1-401-785-3510 (J.B. & L.P.)

Abstract: As part of a science-based conservation strategy for the New England cottontail (NEC), Roger Williams Park Zoo (RWPZ), located in Providence, Rhode Island, participates in a NEC ex situ breeding program in partnership with the NEC Population Working Group. RWPZ's role is to not only breed vulnerable cottontail rabbits for eventual release to bolster wild populations, but also to help foster research on a species that is hard to observe naturally. RWPZ was the first to breed this species in a zoological facility, playing a significant role in the survival of New England's only native rabbit.

Keywords: New England cottontail; ex situ breeding; conservation

Citation: Brown, J.; Puccia, L. Ex Situ Breeding Program with Wild-Caught Founders Provides the Source for Collaborative Effort to Augment Threatened New England Cottontail Populations. *J. Zool. Bot. Gard.* **2022**, 3, 573–580. https://doi.org/10.3390/jzbg3040042

Academic Editors: Ursula Bechert and Debra C. Colodner

Received: 30 September 2022
Accepted: 8 November 2022
Published: 14 November 2022

Publisher's Note: MDPI stays neutral with regard to jurisdictional claims in published maps and institutional affiliations.

1. Introduction

The New England cottontail (NEC) (*Sylvilagus transitionalis*) is a threatened rabbit species (listed as Vulnerable on the IUCN Red List) that used to range throughout much of New England. It is now found only in isolated subpopulations in Connecticut, Massachusetts, Maine, New York, New Hampshire, and Rhode Island. In the latter two states, populations in 2015 were estimated to be less than 180 and less than 100 individuals, respectively [1]. This species requires early successional habitat which is being lost through land development, succession, and suppression of natural disturbances (i.e., wildfires, flooding). NEC numbers have been in decline for decades and, without assistance, the genetic diversity of the remaining isolated populations will continue to decrease as well. They may also be losing the competition for resources with the non-native eastern cottontail (*Sylvilagus floridanus*), which was introduced from Missouri in the 1930′s to benefit hunters [2].

There are many partners in the overall recovery effort working to create or restore early successional habitat on public and private lands in order to expand and reconnect isolated NEC populations [Appendix A]. The missing factor was a source of genetically diverse NECs to (re)populate these areas. In 2010, state biologists reached out to Lou Perrotti, Director of Conservation Programs, and zookeepers at Roger Williams Park Zoo (RWPZ) who had the facilities and expertise to facilitate a breeding program at the zoo. The NEC Population Management Group, comprised of federal and state level wildlife agencies, funding agencies, state universities, and RWPZ, was formed and tasked with the development of a breeding management program for NEC. This program seeks to breed NECs for release into the wild, with the goal of restoring sustainable population numbers.

Within this program, RWPZ is working with students and professors from various universities, providing data for multiple research projects studying the biology and ecology of NEC rabbits. Using this research, RWPZ can fine tune best practices when it comes to NEC breeding management, and field biologists can gain insight into NEC behavior.

2. Ex Situ Breeding Program

Wild rabbits are trapped in Connecticut, Massachusetts, Maine, New Hampshire, and New York, by state wildlife biologists, and brought to the zoo for breeding. These rabbits are called "founders," they are assumed to be unrelated and will form the basis of the population. Upon arrival, their fecal samples are collected and sent out to The University of Rhode Island (URI) for genetic testing to confirm they are NECs and not the similar looking non-native eastern cottontail. Rabbits then undergo a 30-day quarantine. During this time, they are monitored for signs of disease and treated for internal and external parasites. As of 2022 they are also tested for tularemia. Provided the rabbits prove to be healthy NECs, they receive ear tags and microchips for identification, are vaccinated against Rabbit Hemorrhagic Disease Virus (RHDV2), and finally receive a full medical exam before moving into the breeding program.

When the founders are not in a breeding situation, they are housed individually indoors in lab-style suspended caging [Figure 1]. Cages are bedded with timothy hay and furnished with a hide box and pieces of woody browse. Water bottles and grain are mounted on the cage door. Husbandry is performed daily: feces are sifted onto the liner, soiled hay is removed and replaced with new, grain hoppers are refilled, water bottles are changed, and fresh greens are given. Cage liners are replaced every other day. The space is climate controlled and has windows to allow for a more natural photoperiod.

Figure 1. Laboratory-style caging racks. This style of caging, where waste falls through into the liner below, provides a more hygienic space for the rabbits and allows for a much less invasive cleaning style.

Breeding takes place in one of four covered outdoor breeding pens [Figure 2] which are approximately 10′x15′ with a dig proof concrete slab floor and fully meshed sides. Visual barriers are in place between pens to decrease aggression between males. An area of timothy hay is maintained in the center of each pen for both bedding and food. Several hides of assorted sizes are provided for the animals to take shelter together or separately. There are multiple sources for water, grain, and woody browse for them to choose from. Pens are cleaned daily: soiled hay is removed and replaced with new, floors are swept, grain sources replenished, water bottles replaced, and fresh greens are provided. Each pen is fitted with 2 cameras on opposing walls with overlapping fields of view. Camera footage is utilized by students at URI for research.

Figure 2. Outdoor breeding pen. When rabbits move here for breeding, they are moved in their hide boxes which remain in the pen with them to provide additional cover.

Fecal samples are collected from cages or pens when requested for research studies by URI, or sent to State University of New York, College of Environmental Science and Forestry where they are using them to study stress levels in NECs under managed care.

The exact number of males and females in the program each year varies based on the success of the wildlife biologists' trapping efforts. The aim is to have 4–5 males and 13–14 females per season. Rabbit pairings are decided based on maximizing genetic diversity; females are assigned to mate with a different male during each round of breeding and each male is assigned a similar total number of females. NEC pairs are put together in the breeding pens for approximately 48 h unless there are signs of aggression. After this breeding opportunity, males may remain outside to mate again while females are brought back inside to their individual cages.

The females are monitored closely during their birthing window. This window begins 27 days after the first day in the breeding pen and ends 40 days after the last day. If kits are born, noise is kept to a minimum and the cage is not disturbed for 3 days except to deliver fresh browse or to remove any deceased kits: no cleaning takes place at this time in an effort to reduce stress on the dam [Figure 3]. These offspring will be released into the wild and so it is important that they are raised by the dam. The staff strives to provide the best care while maintaining a hands-off approach.

Figure 3. Newborn Kits. A litter of 1-day old kits, eyes shut and ears back. Source: Roger Williams Park Zoo.

Kits remain with the dam until they are weaned at 23 days [Figure 4]. At this time either dam or kits are moved to a new cage and the female can now be bred again. As many breeding rounds as possible are fit into each season, typically starting in February, and ending in August. At the end of the season, some of the adults will be released while those that adapted to the program well and proved to be good breeders may stay at RWPZ for another season. It is important for population genetics to rotate founders so no one rabbit's genes are overrepresented.

After the kits have been weaned, RWPZ veterinarians begin the process of vaccination against Rabbit Hemorrhagic Disease Virus (RHDV2), identifying their sex, and giving them ear tags and microchips. They also clip a tiny portion of one ear to obtain a DNA sample from each animal. Once kits are tagged and identifiable, they are moved outside to one of the empty breeding pens to help them acclimate to living outdoors. Kits remain at RWPZ until the NEC Population Management Working Group decides where they will be released based on their genetic profile. State wildlife biologists will then pick up the rabbits for transport to a hardening pen for further acclimation.

Currently there are two hardening pens, one located in Ninigret National Wildlife Refuge (Charlestown, RI), and the other in Great Bay National Wildlife Refuge (Newington, New Hampshire). The pens are used for a soft-release process specially designed to prepare the young kits for life in the wild. They cover approximately an acre of prime NEC habitat and are predator proof. They have a perimeter fence of hardware cloth extending in an apron below ground level and are covered with nylon netting. Here, the kits learn to forage for themselves and live out in the elements before their final release into the wild. Studies have shown that time spent in pens such as these increases post-release survival [1]. If NECs in the hardening pens are found to need medical attention, they can be brought back to RWPZ's veterinary department.

Figure 4. A dam with her 15-day old kits. The kits will remain with the dam for another 8 days before weaning.

3. Results

Since 2010 there have been 139 founders brought to RWPZ, of those founders 38 died, 84 have been released, and 17 are currently in the breeding program. Over 600 NEC kits have been born at RWPZ and, despite the high mortality rate of newborn rabbits, over 250 of them have been released into the wild [Table 1]. Pre-weaning mortality accounts for the vast majority of losses. The 2022 season is still in progress, but it is already a record year with 94 kits born so far.

Table 1. Total number of NEC kits born at Roger Williams Park Zoo (RWPZ) since 2012, pre-weaning mortality percentage, and number of kits released.

Year	2012	2013	2014	2015	2016	2017	2018	2019	2020	
Number of Kits Born	49	90	83	75	88	67	63	59	9 *	Total = 583
Pre-weaning Mortality %	40.8	52.2	55.4	65.3	64.8	76.1	28.6	28.81	55.5 *	Avg. = 51.9
Number of Kits Released	29	42	37	22	26	21	42	28	4 *	Total = 251

* The breeding program was disrupted/halted during the 2020/2021 seasons due to the COVID-19 pandemic.

Kits from RWPZ have been released on both public and private partner held lands across New Hampshire, Rhode Island, Maine, and Massachusetts. Building on the success at RWPZ, a second breeding program was started at Queens Zoo (Queens, NY) in 2016. Combined numbers from both zoos total close to 1000 NEC births and nearly 400 individuals released into the wild.

In 2012, NEC from RWPZ's breeding program were used in a pilot program seeking to establish a self-sustaining population on uninhabited Patience Island (Portsmouth, RI). The island is free from eastern cottontails and has low predator density compared to the mainland. Over a 5-year period, 73 rabbits born at RWPZ were released onto the island, and in 2017 it was estimated there were at least 200 NEC present. The pilot program was successful, and the colony established on Patience Island now has enough individuals such that this can be a source for starting new populations in and of itself. There are currently 4 self-sustaining populations established with NECs born at RWPZ and Queens Zoo.

4. Conservation Impact

The long-running NEC breeding management program, initiated at RWPZ with additional participation by Queens Zoo, is an important piece of a larger puzzle when it comes to conserving this species. The highly fragmented habitats and populations make it increasingly challenging for the young rabbits to disperse the 1–3 miles they typically travel to breed [2]. Without this option, the population loses the ability to maintain genetic diversity and reduces their ability to adapt to predators, disease, and habitat changes. Increasing the connectivity of these populations will allow them to naturally diversify. As partners in the conservation effort work on this problem through the creation and maintenance of new habitat across New England and New York, RWPZ provides a source of NEC rabbits to colonize these areas. To increase genetic diversity, NECs may also be released into already colonized areas where there is danger of genetic bottlenecking due to low population size.

The genetic profiling done at RWPZ before release is used as a reference for identifying released individuals through DNA analysis of collected feces, a preferred non-invasive monitoring process. In this way RWPZ-bred NEC can be tracked to determine how many survive, if they reproduce, and how far they disperse from the release sites. This method can also be used to monitor population trends such as size and sex ratio to determine if further management action is needed. Data from one 5-year study in New Hampshire has shown that, post release, NEC from the breeding program were able to survive and breed with wild born NEC, increasing genetic diversity [3]. Some rabbits were even able to disperse successfully to an area of suitable habitat 700 m from the release site, showing the potential for repopulating through introduction of individuals from the ex situ breeding program.

Previous and ongoing studies by graduate students at University of Rhode Island (URI) continue to provide insights that help us modify the breeding program and learn about the species. Data obtained from camera footage has been used directly by RWPZ to determine the duration of time breeding pairs spend together and the best breeding pen layouts to reduce aggression and improve mating success. Research from previous years on breeding behavior has implied that pre-copulatory mate choice is occurring, leading to an ongoing mate choice study [4]. Looking forward, the results of this study may mean a novel approach is warranted to allow females to choose more compatible males to breed with, hopefully minimizing breeding violence and increasing birth numbers. It may also impact how biologists view populations in the wild. It would be valuable to know if there is pre-copulatory mate choice occurring in already fragmented wild populations, further diminishing the possible mate choices available.

Since beginning work on conserving this species, the foreign animal disease Rabbit Hemorrhagic Disease Virus (RHDV2) has started to spread throughout the United States from the west coast towards the east. To combat this, all adults brought in and offspring going out are now vaccinated against RHDV2. This process was not easy and RWPZ veterinary staff had to work with the Rhode Island state veterinarian to file a federal application to have the vaccine imported for use. Due to this, all rabbits from Queens Zoo also come to RWPZ for vaccination prior to release. With the first case of RHDV2 reported in Connecticut in September of 2022 (in domestic rabbits) it is more important than ever to make sure that NEC are not wiped out by this new threat [5].

5. Conclusions

Thanks to the work done by RWPZ and all the other partners, this one-time candidate species for protection under the Endangered Species Act no longer meets the listing criteria. Although this is not a trivial achievement, until the habitat and population goals of the conservation strategy are met, RWPZ will continue to assist in increasing population numbers and genetic diversification through the breeding program. The program itself could not be successful without the help of the NEC Population Management Group partners. RWPZ relies on field biologists from other states that still have native populations to provide founders for breeding. Additionally, there is reliance on the university researchers who sift through hundreds of hours of camera footage from breeding pens in order to gain new understanding of NEC behavior. No single effort can save this species on its own, even the NEC Population Management Group is just another piece of the big picture, a subsection of an even larger partner group working to help the NEC populations rebound.

Author Contributions: Writing—original draft, J.B. and L.P. All authors have read and agreed to the published version of the manuscript.

Funding: This conservation program was funded by the Rhode Island Foundation for start-up costs, and the Pittman Robertson Fund and Competitive State Wildlife Grant Program.

Institutional Review Board Statement: This program was approved by the Animal Welfare Committee of Roger Williams Park Zoo on 3 October 2011.

Data Availability Statement: Not applicable.

Acknowledgments: The authors would like to thank Karin Schwartz, Lou Perrotti and Amy Roberts for assisting in gathering data and/or reviewing the manuscript.

Conflicts of Interest: The authors declare no conflict of interest. The funders had no role in the design of the study; in the collection, analyses, or interpretation of data; in the writing of the manuscript; or in the decision to publish the results.

Appendix A

New England Cottontail Rabbit Conservation Partners
CT Department of Energy and Environmental Protection *
Maine Department of Inland Fisheries and Wildlife *
MA Division of Fisheries and Wildlife *
National Fish and Wildlife Foundation *
Natural Resources Conservation Service *
NH Fish and Game *
NY State Department of Environmental Conservation *
RI Department of Environmental Management *
US Fish and Wildlife Service *
University of New Hampshire *
University of Rhode Island *
The State University of New York *
Wildlife Management Institute *
Roger Williams Park Zoo Staff *
WCS Queens Zoo *
U.S. Geological SurveyUSDA Natural Resources Conservation Service
Mashpee Wampanoag Tribe
Lyme Land Conservation Trust
American Forest Foundation
Wells National Estuarine Research Reserve
Woodcock Limited
Audubon Connecticut
Connecticut Audubon Society
Open Space Institute
Audubon New York
Quail Forever
Pheasants Forever
Doris Duke Charitable Foundation

Wildlife Conservation Society
American Bird Conservancy
Quality Deer Management Association
Sustainable Forestry Initiative
White Memorial FoundationRuffed Grouse Society/American Woodcock Society
National Wild Turkey Federation
New England Cottontail Conservation Initiative
Northeast Forest and Fire Management
Lyme Timber Company
Monterey Preservation Land Trust
Narrow River Land Trust
Nantucket Conservation Foundation
Scarborough Land Trust
Avalonia Land Conservancy
Orenda Wildlife Land Trust
The Trustees of Reservations
Berkshire natural Resources Council
York Land Trust
Becket Land Trust
Trust for Public Land
Massachusetts National Guard
Northeast Association of Fish and Wildlife Agencies
* These partners participate in the NEC Population Management Workgroup.

References

1. Litvaitis, J.; Lanier, H.C. *Sylvilagus transitionalis*. The IUCN Red List of Threatened Species 2019: e.T21212A45181534. Available online: https://www.iucnredlist.org/species/21212/45181534#assessment-information (accessed on 26 September 2022).
2. Fuller, S.; Tur, A.; Conservation Strategy for the New England Cottontail (*Sylvilagus transitionalis*). NEC Conservation Strategy Updated 3-11-19. 2012. Available online: www.newenglandcottontail.org (accessed on 20 September 2022).
3. Bauer, M.L.; Ferry, B.; Holman, H.; Kovach, A.I. Monitoring a New England Cottontail Reintroduction with Noninvasive Genetic Sampling. *Wildl. Soc. Bull.* **2020**, *44*, 110–121. [CrossRef]
4. Petit, H.; Richard, J. Behavioral Differences between Successful and Unsuccessful Breeding Pairs of New England Cottontails (*Sylvilagus transitionalis*). Available online: https://web.uri.edu/coastalfellows/behavioral-differences-between-successful-and-unsuccessful-breeding-pairs-of-new-england-cottontails-sylvilagus-transitionalis/ (accessed on 20 September 2022).
5. Connecticut's Official State Website. The Connecticut Department of Agriculture Confirms Incidence of Rabbit Hemorrhagic Disease Virus in Hartford County. Available online: https://portal.ct.gov/DEEP/News-Releases/News-Releases---2022/The-CT-Dept-of-Agriculture-Confirms-Incidence-of-Rabbit-Hemorrhagic-Disease-Virus-in-Hartford-County (accessed on 20 September 2022).

Journal of
Zoological and Botanical Gardens

MDPI

Review

Head-Starting and Conservation of Endangered Timber Rattlesnakes (*Crotalus horridus horridus*) at Roger Williams Park Zoo

Gabriel Montague

Roger Williams Park Zoo, Providence, RI 02907, USA; gmontague@rwpzoo.org

Abstract: The timber rattlesnake (*Crotalus horridus horridus*) is extinct in Rhode Island and Maine with populations in the remaining New England states afforded endangered species status. Lou Perrotti, Director of Conservation and Research at Roger Williams Park Zoo (RWPZ), has long been a champion of these unloved animals in peril and spearheaded a program at the zoo in fall 2010 to work on the recovery of this endangered snake species. Partnering with multiple state agencies was required to begin saving this maligned native species, which had seen massive population reductions. The program began with accepting adults of varying size and sex suffering from skin lesions believed to be Snake Fungal Disease from multiple New England states. Depending on the severity of the infection, the animals were treated and then released. As the program evolved, it became a goal to not only treat affected adults and determine the overall health of declining New England populations but to begin a head-start program with one of the state conservation partners. Head-starting refers to when neonates are either born in a zoological facility or captured in the wild and raised under managed care until a desired size is reached. They are then released back to the wild, giving them a better chance for survival. The area where the snakes are kept at the zoo allows for temperature manipulation to simulate a natural temperature change and allow for the brumation of individuals. Once the appropriate size is reached, a radio transmitter is surgically implanted to allow radio telemetry tracking after release. The head-starting strategy has been a success, with individuals being found years later, suggesting they are surviving in the wild. Due to the sensitive nature of this program, some data and names of partners have been purposely omitted.

Keywords: rattlesnake; head-start; Conservation; zoo; endangered species

Citation: Montague, G. Head-Starting and Conservation of Endangered Timber Rattlesnakes (*Crotalus horridus horridus*) at Roger Williams Park Zoo. *J. Zool. Bot. Gard.* **2022**, *3*, 581–585. https://doi.org/10.3390/jzbg3040043

Academic Editors: Ursula Bechert and Debra C. Colodner

Received: 30 September 2022
Accepted: 9 November 2022
Published: 14 November 2022

1. Introduction

The timber rattlesnake (*Crotalus horridus horridus*) is an iconic American animal appearing on the famous Gadsden Flag [1] and other patriotic symbols throughout the history of the United States. Despite being a patriotic symbol for the United States of America, the animal has been mercilessly persecuted by humans for being a venomous snake. Timber rattlesnakes are relatively large snakes, ranging from 35 to 60 inches in length. They are a pit viper (family Viperidae, subfamily Crotalinae) with a prominent tail rattle and heavily keeled scales. Hatred and ignorance have had a devastating effect on the timber rattlesnakes' population. New England is at the northern extant of the timber rattlesnakes' native range, so naturally, life in this region is a little more difficult due to the northern climate, even without any additional external pressures. The past two hundred years have seen bounties for dead animals, extreme habitat destruction and degradation, road mortality, and the recent emergence of Snake Fungal Disease (SFD). SFD is caused by *Ophidiomyces ophiodiicola*, a fungus that was recently split from a group of fungi long referred to as the *Chrysosporium* anamorph of *Nannizziopsis vriesii* species complex (CANV). A primary infection by *O. ophiodiicola* may initiate a series of events that lead to the death of the host snake. It is also plausible that in some instances, the fungal colonization of tissues occurs because the host's health is already compromised [2].

Timber rattlesnakes have very specific habitat requirements, with pine and mixed deciduous forests nearby to mountainous areas and access to deep underground dens to allow for brumation (a period of dormancy over the winter months). The underground dens are usually near a southern facing talus (loose rock) slope with little vegetative cover. Other specific habitat requirements include birthing areas (rattlesnakes give birth to live young) usually protected by rock ledges, basking areas, and even trees to climb [3]. Unlike true hibernation, they have periods of activity in the brumation period. Entering their dens in mid to late October, they occasionally move around in the dens and emerge in April, basking frequently until leaving to forage for the summer months. In addition, long time spans are required to attain sexual maturity, long inter-birthing intervals occur in individual females, and a small litter size make this rattlesnake vulnerable to a plethora of threats [4]. Juvenile rattlesnakes face many difficulties as they grow, as they are food for many predators including Virginia opossums, raptors, bobcats, and even other snakes. During a previous study with timber rattlesnakes, it was determined that for snakes in the first two age groups (ages 1 yr and 2–4 yr), survival tended to decline over time for both color variations (darker scalation or more yellowish), while for adult snakes (5 yr and older), survival was static or even slightly improved when raised in managed care [5].

The basic goal of the SFD study was to take in animals suffering from various stages of perceived SFD infections and, with various treatments, support the animals in achieving sufficient health so they could be released back to their home range. For the head-starting program, the neonates were raised in managed care until they attained a weight heavy enough (~300 g) to allow for a small radio transmitter to be surgically implanted. This process usually takes two and a half years of being in managed care and this size also makes it more likely the animals will survive into adulthood as they are larger than a wild counterpart of the same age.

2. Materials and Methods

2.1. *Rehabilitation and Treatment of SFD*

In the beginning of the project (October 2010 until 2016), the main goal was the determination of the overall health of animals suffering from skin lesions believed to be caused by SFD. To begin the rehabilitation and treatment of adult individuals suffering from severe fungal infections, a suitable space was identified for its ability to be temperature controlled with air conditioning in the warmer months and with heat in the winter months. One incredibly important detail of this room was the ability to drop the temperature and photoperiod low enough in the late fall and winter to simulate what the snakes would experience in the wild as they entered brumation. A team of keepers and veterinary technicians capable of working with venomous snakes was identified by the Director of Conservation and Lead Keeper. Since SFD is a highly contagious pathogen, a basement room was chosen since it was not close to any of the zoo's collection snakes. This room was treated as a quarantined space. A footbath containing disinfectant was used before entering and leaving the room. This served the dual purpose of limiting exposure of SFD through the tracking of fungus on footwear to the zoo's snake collection. All servicing of the zoo's collection snakes was completed before entering the quarantine area and hands were always washed prior to and after doing any work with the conservation snakes.

2.1.1. Husbandry

For each snake, Vision brand caging (models 211 and 322) was set up along with a water bowl, small heating pad on rheostat, appropriately sized artificial hide, fluorescent lighting (on a timer), and newspaper as a substrate. The paper was chosen over a more natural substrate for its ease of cleaning, which was very important for the animals that required SFD treatment. Keepers always worked in teams of two. This was to ensure safety and, in the event of a keeper being bitten by a rattlesnake, the snake bite protocol could be immediately initiated. In both mid spring and late fall, temperature and light cycles were manipulated to simulate the seasonal changes they would encounter in the wild. In the late fall, roughly two weeks after the last feed, the heat pads were unplugged, and the room temperature was dropped by 5 °F every five days until the ambient temperature was 47–52 °F. The light cycle would be reduced on the days the temperature was dropped beginning with a summer cycle of 14 h light and 10 h dark. When the last cycle of 8 h light and 16 h dark was achieved, the next step was to turn the lights off completely, only turning on a room light to perform daily checks while the animals were in hibernation. This process was typically reversed beginning in late March; although, there were years when warm spring temperatures forced us to begin the process earlier. Once the snakes had been at an ambient temperature of 75–78 °F with a full photoperiod for a week, they were offered their first meal. Throughout their captivity, the snakes were offered an appropriately sized food item, usually a small rat or large mouse, and fed on a bimonthly schedule. A few of the snakes with severe cases of SFD required surgery under anesthesia for debriding of the affected area. One individual needed to recover from a perceived predator attack that it survived in the wild.

2.1.2. Medical Assessment and Treatment

To test for the prevalence of SFD, biopsies were taken from the scales of snakes presenting symptoms of a fungal dermatitis, and blood samples were taken to help determine the overall health of affected individuals. For the actual treatment of the rattlesnakes, the Veterinary Department determined what medical treatment was needed besides antifungal (Voriconazole) drugs. Often, secondary infections were present in the animals and the antibiotic ceftazidime along with subcutaneous fluids were used to treat these infections. When the rattlesnakes needed to be handled for a medical treatment or checkup, a keeper removed the snake from its habitat with an appropriately sized snake hook and placed the animal on the floor along a wall. The keeper would then attempt to coax the snake into an appropriately sized transparent plastic tube. Making sure the proper tube was selected was a very important part of this process. If the tube was too large, the snake could turn around and envenomate the keeper. If it were too small, the snake could become stuck and be difficult to safely remove. The keepers often had to switch the hook and tube between hands while attempting to coax the snake into the tube. The snakes, being wild, were often reluctant to take part in this exercise. When roughly 35% of the snake was inside the tube, the keeper would make a single swift movement of setting down the hook and grabbing the snake right where the tube and snake met. Once the snake had been secured in the tube, its body could be picked up with the keeper's free hand so that it was fully supported. After the required treatment was given, the snake was returned to its habitat. Usually by this point, the snake's mood had soured due to being handled and treated so it was important to be as safe as possible. The snake would, therefore, be placed in its habitat while still secured in the tube. In one fluid motion, the keeper would release the snake and tube while simultaneously using the hand holding the upper part of the tube to push the snake away from the front of the enclosure. This motion, coupled with the snake's natural inclination to back out of the tube, allowed the second keeper to close the door as the main keeper backed up.

2.2. *Head-Starting Neonates for Release to the Wild*

The project shifted from treating infected adults to head-starting in 2016. The husbandry processes remained the same (temperature and lighting), but the caging had to be quite different due to the snakes' small size (usually between 18 g–25 g). The neonate snakes were kept in rack style caging (individual tubs had measurements of 36″L × 16″W × 12″T) with a locking lid, and these contained a small water bowl and an artificial hide. A large fluorescent tube light was placed in front of the whole rack of caging.

The neonates were acquired in two separate ways. In the first, state biologists would catch a known gravid female who would give birth at the zoo. In the second, the neonates were captured in the wild at known birthing sites. Neonates were weighed, placed in their rack, and after their first shed (usually within a week), they were offered their first meal. Although the adults were fed every other week to every third week, depending on body condition, neonates were offered food once a week. Most neonates readily accepted thawed fuzzy mice as their first meal, with a few exceptions needing to be started on thawed pinky mice. The juvenile snakes were kept until they reached a size of around 300 g when they could be surgically implanted with a small radio transmitter allowing biologists to track them when they were released. The growth process usually took around two and a half years, and they were released where they originally came from in mid to late spring depending on the weather. All snakes were also implanted with a Passive Integrated Transponder (PIT) so that in the event of the radio transmitter failure they could still be identified.

3. Results

Between 2010 and 2016, 96.1% of adult rattlesnakes brought in for the study and treatment of SFD were returned to the wild after being cleared of medical issues. After release, the animals were monitored by state biologists and seemed to do well, being observed year after year; however, the fungus was a continuing threat since it is present in their environment. The 3.9% of adults that were not able to be released succumbed to the effects of severe SFD and secondary infections. The head-start program that began in 2011, and is ongoing, has a success rate of 91.6% of individuals being released back to their wild habitat. The 8.4% of juveniles that were not released include the animals currently in managed care and the small percentage of animals lost due to a diagnosis of failure to thrive. Through radio tracking, it has been noted that released animals are currently thriving and successfully hibernating, which can be one of the biggest hurdles for an animal raised under human care. However, to be successful, released animals must demonstrate competency in the wild such that they can select appropriate resources, respond correctly to seasonal environmental cues, avoid predators, and ultimately grow, survive, and reproduce [6]. Due to the sensitive nature of this information, some data and locations have been purposely omitted.

4. Discussion

While the timber rattlesnake faces many threats from both natural and anthropogenic sources, rehabilitation and head-starting programs such as this can continue to provide some relief to struggling wild populations. The juvenile snakes, when released, are larger than a wild counterpart of the same age and the threats they face would be the same as an adult snake. Timber rattlesnakes are a vital part of the ecosystem in New England, being predators of small mammals that are primary vectors for Lyme disease. In a study conducted in the George Washington National Forest, it was observed that 87% of snakes had consumed small mammals and 13% had consumed birds (Uhler et al., 1939) [7]. They also are prey for many other native species such as bobcat, red tailed hawk, Virginia opossum and fox to name a few. Sadly, however, their value to the ecosystem is often overlooked and they are not beloved. They are victims of society's perpetuated intolerance and commonly encouraged phobias for serpents, especially venomous ones. One of the most important things timber rattlesnakes have going for them is secrecy, as very few people realize they even live here, and fewer are lucky enough to see them in the wild. There are many possible directions for future research regarding head-starting rattlesnakes, including reproductive studies of head-started snakes compared to wild snakes, long term studies of growth rates, population stability in head-started populations compared to wild snakes and its efficacy in helping to maintain and even bolster declining wild populations. Natural mortality rates for timber rattlesnakes in their first two years of life are as high as 66%, due to failure to procure food, predation, and failure to find proper hibernacula [7]. For a species that is barely clinging to life in New England, the importance of a program such as the one here at RWPZ cannot be overstated.

Funding: Funding for the SFD study was provided by a USFWS Regional Conservation Needs grant (RCN) number 2012-02 and grant amount was not to exceed $81,148.71. The funding for the head-start program, treatment and husbandry of all snakes is provided in house by RWPZ.

Institutional Review Board Statement: The animal study protocol was approved by veterinary department and senior management at Roger Williams Park Zoo.

Informed Consent Statement: Not applicable.

Acknowledgments: The author would like to thank the following: Lou Perrotti, Director of Conservation and Research at Roger Williams Park Zoo and Carousel Village and his vision and drive to work with a controversial species; the multiple New England state conservation departments that are willing to protect a species that suffers from such poor public perception, which can hamper the conservation process; my wonderful wife for taking the time to edit the review; and lastly, I would like to thank Karin Schwartz for being patient and her endless professional assistance in writing this paper.

Conflicts of Interest: The authors declare no conflict of interest.

References

1. Revolutionary War and Beyond. Gadsden Flag. Available online: revolutionary-war-and-beyond.com (accessed on 17 September 2022).
2. Lorch, J.M.; Lankton, J.; Werner, K.; Falendysz, E.A.; McCurley, K.; Blehert, D.S. Experimental infection of snakes with Ophidiomyces ophiodiicola causes pathological changes that typify snake fungal disease. *mBio* **2015**, *6*, e1534-15. Available online: https://journals.asm.org/doi/epub/10.1128/mBio.01534-15 (accessed on 15 September 2022). [CrossRef] [PubMed]
3. Massachusetts Division of Fisheries & Wildlife. Living with Wildlife: Timber Rattlesnakes in Massachusetts. Available online: https://www.mass.gov/doc/living-with-wildlife-timber-rattlesnakes-in-massachusetts (accessed on 24 September 2022).
4. Brown, W.S. Biology, Status, and Management of the Timber Rattlesnake (*Crotalus horridus*): A Guide for Conservation. *Herpetol. Circ.* **1993**, *22*, 1–78.
5. Brown, W.S.; Kéry, M.; Hines, J.E. Survival of timber rattlesnakes (Crotalus horridus) estimated by capture-recapture models in relation to age, sex, color morph, time, and birthplace. *Copeia* **2007**, *3*, 656–671. [CrossRef]
6. Alberts, A.C. Behavioural considerations for headstarting as a conservation strategy for endangered Caribbean Rock Iguanas. *Appl. Anim. Behav. Sci.* **2007**, *102*, 380–391. [CrossRef]
7. Virginia Herpetological Society. Timber Rattlesnake (*Crotalus horridus*). Available online: www.virginiaherpetologicalsociety.com/reptiles/snakes/timber-rattlesnake/timber_rattlesnake1.php (accessed on 2 November 2022).

Journal of
Zoological and Botanical Gardens

MDPI

Review

Over 25 Years of Partnering to Conserve Chiricahua Leopard Frogs (*Rana chiricahuensis*) in Arizona, Combining Ex Situ and In Situ Strategies

Tara R. Harris [1,*], Whitney L. Heuring [1], Ruth A. Allard [1], Audrey K. Owens [2], Shaula Hedwall [3], Cat Crawford [4] and Christina Akins [5]

1 Conservation and Science Department, Arizona Center for Nature Conservation/Phoenix Zoo, Phoenix, AZ 85008, USA
2 Ranid Frogs Project, Arizona Game and Fish Department, Phoenix, AZ 85086, USA
3 Southwest Forest Complex, U.S. Fish and Wildlife Service, Flagstaff, AZ 86001, USA
4 Arizona Ecological Services, U.S. Fish and Wildlife Service, Tucson, AZ 85745, USA
5 Payson and Pleasant Valley Ranger Districts, Tonto National Forest, U.S. Forest Service, Payson, AZ 85541, USA
* Correspondence: tharris@phoenixzoo.org

Abstract: The Phoenix Zoo has partnered with US Fish and Wildlife Service, Arizona Game and Fish Department, US Forest Service, and other organizations for more than 25 years to help recover Chiricahua leopard frogs (*Rana* [=*Lithobates*] *chiricahuensis*) in Arizona, USA. This federally threatened species faces declines due to habitat loss and degradation, long-term drought, disease, and invasive species. Over 26,000 larvae, froglets, and adults, as well as 26 egg masses produced by adults held at the Phoenix Zoo have been released to the wild, augmenting and/or re-establishing wild populations. Chiricahua leopard frog-occupied sites in Arizona have increased from 38 in 2007, when the species' recovery plan was published, to a high of 155 in the last five years, as a result of ex situ and in situ conservation efforts. As one of the longest-running programs of its kind in the United States, communication among partners has been key to sustaining it. Recovery strategies and complex decisions are made as a team and we have worked through numerous management challenges together. Though Chiricahua leopard frogs still face significant threats and a long road to recovery, this program serves as a strong example of the positive effects of conservation partnerships for native wildlife.

Keywords: conservation partnerships; amphibian; translocation; release; head-starting; reintroduction; breeding

Citation: Harris, T.R.; Heuring, W.L.; Allard, R.A.; Owens, A.K.; Hedwall, S.; Crawford, C.; Akins, C. Over 25 Years of Partnering to Conserve Chiricahua Leopard Frogs (*Rana chiricahuensis*) in Arizona, Combining Ex Situ and In Situ Strategies. *J. Zool. Bot. Gard.* **2022**, *3*, 532–544. https://doi.org/10.3390/jzbg3040039

Academic Editors: Ursula Bechert and Debra C. Colodner

Received: 27 September 2022
Accepted: 24 October 2022
Published: 28 October 2022

1. The Phoenix Zoo's Legacy of Native Species Conservation

The Arizona Center for Nature Conservation/Phoenix Zoo (hereafter, Phoenix Zoo/Zoo) is one of the largest private, nonprofit zoological facilities in the United States. Located in Phoenix, Arizona, the Zoo opened in November 1962 and is accredited by the Association of Zoos and Aquariums (AZA) [1]. In addition to exhibits and experiences open to the public, the Zoo is also the site of the Arthur L. and Elaine V. Johnson Native Species Conservation Center (Johnson Center), where staff work with ten species of conservation concern in Arizona in collaboration with the US Fish and Wildlife Service (USFWS), Arizona Game and Fish Department (AZGFD), US Forest Service (USFS), US Bureau of Land Management, US Geological Survey, private land managers, universities, other AZA-accredited zoos and conservation organizations, and more. These efforts include propagation-for-release programs to augment wild populations and scientific research that helps inform species management plans at the Zoo and in the field [1]. The Phoenix Zoo also supports international wildlife and habitat conservation projects, some led by Zoo scientists and others directed and managed by individuals and organizations based in the field [1]. None of this work would be possible without continued support of the Zoo's board of trustees and staff leadership, which have prioritized conservation funding even during times of economic uncertainty. In addition, the Zoo has benefited from generous contributions from individual donors, grants from foundations and federal/state agencies, and corporate support for its mission, which specifically calls out stewardship and conservation of wildlife and their habitats. While all modern, professionally managed zoological parks contribute to wildlife conservation as a requirement of AZA accreditation [2], the Phoenix Zoo is especially recognized for its long-term efforts to help recover species native to its home state [1]. Initiated in 1995, the Zoo's Chiricahua leopard frog program, in partnership with AZGFD, USFWS, and USFS, is one of its flagship conservation efforts [1].

2. Chiricahua Leopard Frogs: Background

The Chiricahua leopard frog (*Rana* [=*Lithobates*] *chiricahuensis*) is listed as a federally threatened species in the United States [3], as amenazada (=threatened) on the Lista de Especies en Riesgo of Mexico [4], and as vulnerable on the International Union for Conservation of Nature Red List [5]. The species is native to a variety of permanent and semi-permanent aquatic systems in Arizona and New Mexico, USA, as well as Sonora, Chihuahua, and Durango, Mexico [3,6–8]. Found in areas ranging from montane pine woodlands to lowland grasslands, and from pristine spring-fed pools to streams and earthen cattle tanks, these frogs are habitat generalists (reviewed in [3,6,7]). Habitat loss and degradation, long-term drought, disease, and the establishment of invasive non-native predators such as bullfrogs and crayfish in many permanent waters within the species' range led to decline of Chiricahua leopard frog populations, which are now largely restricted to artificial aquatic systems and natural systems that lack invasive predators (reviewed in [3,6–8]).

Chiricahua leopard frogs were listed as threatened under the U.S. Endangered Species Act in 2002 after USFWS determined the species was absent from more than 75 percent of historical sites and that significant threats to the species will continue for the remaining small and scattered populations [6]. The USFWS organized a formal recovery team tasked with developing a recovery plan, completed in 2007 [7], for the species. To be considered for delisting, each of eight geographically defined recovery units across the species' range must meet criteria regarding long-term persistence of metapopulations and isolated populations, protection and management of aquatic breeding habitats and the areas connecting them, and reduction or elimination of threats to the species and other causes of population decline [7]. The recovery plan also identified management areas within each recovery unit where potential for successful recovery actions is greatest [7].

Ex situ conservation strategies for Chiricahua leopard frogs began well before the development of the species' recovery plan and were incorporated into the plan as a critical tool for accomplishing recovery actions, including reestablishment of populations at formerly occupied sites, augmenting populations at occupied sites, and temporary rescue of frogs facing imminent acute threats followed by repatriation after threats have been abated [7]. Today, the ex situ conservation program for Chiricahua leopard frogs is one of the longest-standing and largest programs of its kind in the United States, providing more than 26,000 individuals for release to the wild from the Phoenix Zoo alone (see below), and is an important tool for working towards recovery of the species [9,10].

3. History of the Phoenix Zoo's Involvement in Chiricahua Leopard Frog Conservation

In 1995, the Phoenix Zoo was invited to collaborate on leopard frog conservation with a large partnership that included state and federal wildlife agencies, zoos, nongovernmental organizations, corporations, universities, ranchers, and private landowners. In the early years of this work, egg masses were brought in from the wild and reared in small aquaria and plastic kiddie pools in a curator's office, with all resulting head-started frogs released before winter to augment wild populations [11]. Over time, the Zoo dedicated additional physical resources to the program, detailed later in this paper. The program initially relied on a volunteer "Tadpole Taskforce" whose members were responsible for daily husbandry of animals, under oversight from the Zoo's Living Collections team. In 2008, the Zoo established its Conservation and Science Department, and began adding staff dedicated specifically to management and care of native species conservation programs. Soon after, the Tadpole Taskforce was disbanded, and paid Zoo technicians became solely responsible for daily husbandry. Throughout the program, Zoo scientists have joined state and federal agency biologists on field surveys and monitoring excursions and conducted some surveys on their own, further supporting recovery efforts and increasing Zoo biologists' connection to the work in the species' native range [12].

4. Evolution of Ex Situ Conservation Facilities and Strategies

The Phoenix Zoo has contributed to the recovery of Chiricahua leopard frogs in Arizona using two main ex situ conservation strategies—head-starting and breeding. When head-starting, the Zoo has typically received wild egg masses collected by conservation partners and reared them to late-stage larvae or juvenile frogs to increase survivorship beyond what would be likely in the wild. These individuals are then released into wild sites within their respective recovery unit. The Zoo has also bred adult frogs and provided egg masses for release to the wild. These conservation strategies are used to start new populations or augment existing populations in areas with suitable habitat and low levels of threat. During annual recovery meetings, partners identify existing, restored, or created habitats for future introduction or augmentation. Primary factors influencing site selection include water permanency, proximity to extant sites or metapopulations, existing threats, and genetics.

Each strategy has required different types of facilities and management, which evolved over the years as the species' recovery program needs changed. The first facility at the Phoenix Zoo built for Chiricahua leopard frog rearing was the Montane Anuran Conservation Center [12]. It was constructed from two insulated cargo containers and equipped with air conditioning units, lighting, and aquaculture tubs that served as a small rearing facility. This facility was used to head-start over 5000 Chiricahua leopard frogs between 1997 and 2007. Zoo staff also repurposed an existing outdoor space, known as the Lower Anuran Conservation Center (LACC), that had been part of fish hatchery operations on site prior to the Zoo's opening in 1962. The LACC was an L-shaped concrete tank with mesh roof and side panels that held leopard frogs from 1996 until it was decommissioned in 2013 due to new construction. The LACC allowed for overwintering of adult frogs and production of egg masses on Zoo grounds.

In 2007, the Phoenix Zoo opened the Arthur L. and Elaine V. Johnson Foundation Conservation Center (Johnson Center), a permanent facility with space for breeding and rearing multiple species, including a ~850 ft2 lab dedicated to Chiricahua leopard frog rearing (Figure 1). Opening the Johnson Center marked a pivotal point in the Chiricahua leopard frog program, as it allowed for multiple populations of frogs to be reared simultaneously in biosecure facilities and included a separate space for quarantining frogs coming in from the field [12]. The Johnson Center magnified the Phoenix Zoo's recovery contribution by increasing the number of individual Chiricahua leopard frogs reared and released into the wild.

Figure 1. Chiricahua leopard frog head-starting lab in the Phoenix Zoo's Johnson Center.

Another expansion occurred in 2010 when the Johnson Conservation Center Ranaria Complex (Ranaria Complex) was constructed, providing fully contained semi-natural habitats for housing and rearing Chiricahua leopard frogs and other aquatic species outdoors (Figure 2). This complex has been renovated over the years to accommodate the changing needs of the Phoenix Zoo's native species programs and currently contains 12 enclosures, six of which now house Chiricahua leopard frogs. Each enclosure has a pond with a filtration system surrounded by land, with live terrestrial and aquatic plants providing varied cover. This outdoor holding space mimics the natural environment and helps promote foraging and cryptic behaviors in the frogs, as well as exposes them to seasonal temperature and light fluctuations.

Figure 2. Phoenix Zoo Ranaria Complex habitats for Chiricahua leopard frog rearing and breeding.

One benefit of the Ranaria Complex has been the ability to house male and female frogs separately. This increased capacity allows the Zoo to hold adult populations year-round and pair wild-sourced individuals representing remaining genetic lineages for breeding when timing is right to produce egg masses. Egg masses are then translocated to wild sites or brought into the Johnson Center for rearing. The creation of additional frog enclosures in the Ranaria Complex allowed the Zoo to move from focusing efforts on rearing frogs from one population in one recovery unit at a time to rearing multiple populations in different recovery units or different management areas within a recovery unit simultaneously. Partners working to manage wild Chiricahua leopard frogs rely heavily on the Ranaria Complex as a source for reestablishing frog populations using egg masses produced by adult breeders when conditions in the wild lead to sudden population declines. When biologists conduct emergency salvages at drying sites, the Ranaria Complex has also acted as a temporary refugium until water naturally returns to these sites.

As the facilities at the Phoenix Zoo have evolved through the years, larval head-starting and rearing strategies have also changed. In early years, the focus was on producing large numbers of larvae [12]. More recently, efforts have shifted focus to ensure animals produced are large and healthy at the time of release. For many years, all individuals were released to the wild by fall, but the addition of the Ranaria Complex allowed the Phoenix Zoo to start overwintering some larvae and juvenile frogs. This strategy, followed by a springtime release of larger frogs, may increase their chances of survival after translocation by providing frogs more acclimation time prior to winter when they are vulnerable to chytridiomycosis, the disease caused by the fungal pathogen *Batrachochytrium dendrobatidis* (*Bd*). This longer "growing season" may also provide released frogs a chance to breed prior to their first winter in the wild, which could increase population persistence through overwintering larvae at sites where adults and juveniles are susceptible to chytridiomycosis in winter [13,14]. This is of particular importance for higher-elevation populations of Chiricahua leopard frogs, which appear severely limited by localized die-offs from chytridiomycosis.

Husbandry protocols adhere to the Phoenix Zoo's Standard Operating Procedure (SOP) for Chiricahua leopard frogs, informed by the species' recovery plan [7], which outlines daily and monthly routines, capture and handling protocols, and wellness evaluations. The SOP is a flexible document that is updated as protocols change. Protocols have evolved from rearing low numbers of larvae in numerous small tanks to rearing larger groups of larvae in bigger tanks to optimize staff time and resources. To provide a better environment for larvae to grow, the Phoenix Zoo and partners also invested in improving the quality of water used in the frog head-starting lab. Shifting away from dechlorinated tap water, this lab now relies on a reverse osmosis deionized water system which produces pure water that is then reconstituted, resulting in water with a concentration of solutes that is isotonic with the internal concentration of amphibians. Housing and rearing of Chiricahua leopard frogs at the Zoo has also allowed staff to collect data on body size and life stage to determine growth and development rates and improve the species' management (Phoenix Zoo, unpublished data). To help improve nutrient absorption and body condition of larvae and juveniles, the Zoo has modified lighting and diet offerings. Zoo scientists continue to evaluate husbandry and head-starting strategies with the goal of producing healthy individuals that will thrive in the wild.

5. Conservation Successes at the Zoo and in the Field

Over the years, the Phoenix Zoo has been successful rearing larvae and juveniles for release in Arizona. Mean survival for individuals from newly hatched larval stage to release to the wild is 70.1% (Table S1; range: 43.9–97.2%; years: 2009–2020, when consistent data collection methods allowed for comparison). Survival varies among years, with no trend in survival of larvae over time ($n = 16$; $r^2 = 0.0016$; $p = 0.882$). When multiple egg masses were reared simultaneously in the same lab, suvival of larvae varied by as much as 45.2% between egg masses (Table S1), suggesting there are factors other than rearing environment (e.g., genetics and pathogens) playing a role in survivorship. Though survivorship of early life stages is largely unknown for wild Chiricahua leopard frogs, overall survivorship is low [7]. Given that, for most amphibians with indirect development, early life stage mortality is very high (>90% [15]), head-starting Chiricahua leopard frogs past these stages and releasing large numbers of individuals, as is possible through the Zoo's facilities, can be an important strategy for improving survivorship [7,10].

From 1995–2020, 179 releases of Phoenix Zoo-reared leopard frogs have taken place in Arizona, totaling 26,821 individual larvae and frogs (Figure 3; Table S2) and 26 egg masses, which roughly equates to 20,000 eggs (assuming ~750 eggs per egg mass). The number of individuals produced and released increased after the opening of the Johnson Center in 2007.

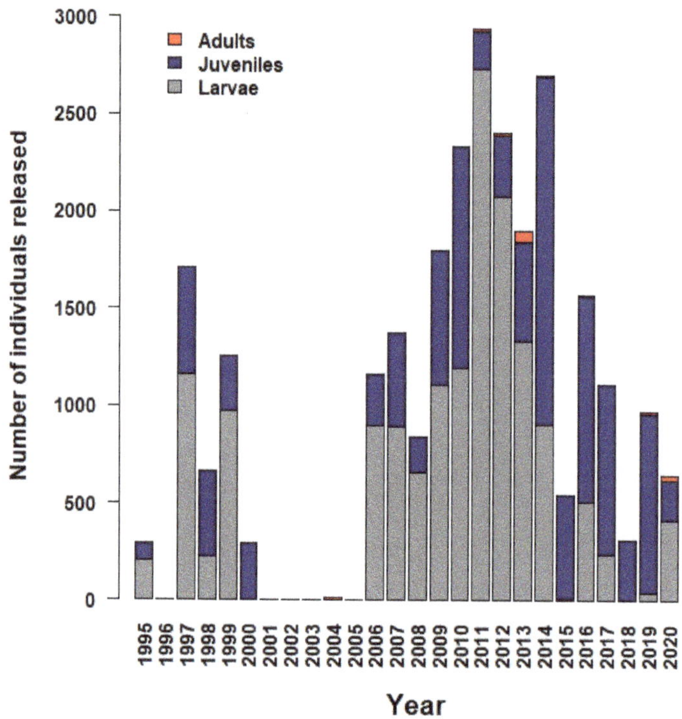

Figure 3. The number of Chiricahua leopard frogs released from 1995–2020 after rearing and/or housing at the Phoenix Zoo. A total of 26,821 individuals (15,562 larvae, 11,119 juvenile frogs, and 140 adult frogs) were released during this timeframe. The Zoo's program involvement was reduced in 2001–2005 due to staffing changes.

Wild Chiricahua leopard frog population status, including the success of releases, is gauged primarily using visual encounter surveys. Current monitoring in Arizona includes visual encounter surveys at sites, where surveyors record presence or absence of frogs and collect data on presence of threats, habitat conditions, evidence of breeding, and more. Surveys sites are selected to document (1) continued persistence of Chiricahua leopard frogs or habitat; (2) persistence of frogs following a recent release; (3) dispersal events; and (4) habitat suitability [9]. In recent years, sites have typically been surveyed between one and three times annually—in spring to detect overwinter survival and breeding, early summer prior to the monsoon to gauge water permanency, and post-monsoon (September–October) to detect dispersal and breeding.

Completion of the Chiricahua Leopard Frog Recovery Plan in 2007 [7] marked a turning point in the conservation of the frog by providing a blueprint to work towards recovery of the species. In 2007, there were only 38 known sites occupied by Chiricahua leopard frogs in Arizona. While the Recovery Plan outlined surveying protocols, there is still a need for a scientifically rigorgous, long-term monitoring program across the species' range that would allow for inferences about the species' status over time. Current population assessments in Arizona are made annually based on the number of sites in which we document frogs, breeding, and a robust population (as defined in the Recovery Plan) [7,9]. The number of occupied sites in Arizona fluctuates annually based on persistence of frog populations and annual factors including precipitation and survey effort; between 2016 and 2021 the number of occupied sites in Arizona was as high as 155—a four-fold increase relative to 2007 [9]. This progress is due to a multi-pronged recovery approach that includes translocations, bullfrog control, habitat restoration, conservation agreements with private landowners, building support through outreach, and application of research and monitoring through adaptive management [9].

A recent analysis of 25 years of Chiricahua leopard frog translocation events [10], with most releases involving animals from the Phoenix Zoo, highlights the value of translocations and provides useful information for adaptive management. Translocations from captive, semi-captive, and wild source populations were all associated with increased probability of release site population persistence, i.e., the proportion of years that sites were estimated to be occupied by Chiricahua leopard frogs [10]. Of the various life stages translocated to wild sites, larval releases were associated with the largest increases in population persistence [10]. Persistence probability also increased as numbers of translocation events at a site increased, with two or more translocations at a site associated with four or more years' increase in the site's predicted occupancy [10]. The combination of several stocking events with large numbers of larvae into neighboring lentic sites that lack vertebrate predators maximized translocation success [10]. Stocking large numbers of Chiricahua leopard frog larvae depends on animals produced en masse at the Phoenix Zoo and cannot be replicated through wild to wild releases.

6. Evolution and Maintenance of Partnerships

Species recovery is typically a complicated, lengthy process with successes and set-backs [16–21]. Strong partnerships are critical to keeping recovery programs moving forward in the face of what can sometimes feel like insurmountable challenges, such as long-term drought and intractable disease in the case of Chiricahua leopard frogs. When partners support one another and work together to devise new ideas and learn from both successes and failures, a recovery program is greatly strengthened.

Communication has been key to the longevity of the Chiricahua leopard frog conservation partnerships in Arizona. Specifically, we credit our ability to have difficult conversations and make hard choices and program improvements together as a team, putting the conservation of frogs first and working toward a common goal. With commitment to the partnership becoming institutionalized over time, we have also been able to maintain momentum even as staff changes occurred in key roles at partner agencies and organizations. Once the recovery plan was completed, the formal recovery team for the Chiricahua leopard frog was replaced by steering committees in each U.S. state where the frog occurs as well as regional work groups. High level priorities, such as new research, adaptive management, range-wide threat abatement, and other range-wide recovery activities are discussed annually at steering committee meetings in Arizona and New Mexico. The recovery plan guided regional work groups to implement recovery at a local level based on the eight recovery units and their associated management areas. As a result, eight local recovery groups across Arizona meet annually to prioritize and implement short- and long-term conservation and recovery actions in each management area or recovery unit.

Furthermore, a key to the success of the partnership in Arizona and beyond is that both in situ and ex situ conservation strategies have been integrated since the beginning and formalized in conservation plans—especially the Chiricahua Leopard Frog Recovery Plan [7,22]. By taking a "One Plan Approach" to species conservation planning [23,24] rather than planning separately for in situ and ex situ conservation efforts, we have ensured that we are all working together toward common goals and objectives. Furthermore, such integrated planning benefits the partnership by demonstrating respect for the different but complementary roles and expertise that in situ and ex situ conservation partners contribute to species recovery.

Over the years, the Phoenix Zoo has become more involved in broader recovery efforts for Chiricahua leopard frogs, including participating to a greater extent in planning and monitoring efforts. The bulk of the Zoo's involvement has focused on work in Arizona, with communication and collaboration with colleagues working in New Mexico as requested. An annual agency-sponsored Chiricahua leopard frog certification workshop and associated field survey training has allowed Zoo staff and numerous others to gain knowledge and skills needed to help monitor wild populations. Furthermore, important for the partnership and recovery program has been the Zoo's willingness and ability to pivot quickly and be flexible to the needs of the species and those who manage them, such as taking in additional animals, sometimes from new recovery units or management areas of Chiricahua leopard frogs on short notice. Without this versatility, populations or genetic lineages of frogs potentially would have been lost.

7. Challenges Faced—Example 1: Deciding to Cross Genetic Lineages

Across recovery programs for imperiled species, managers and conservation partners sometimes face the difficult decision of whether to artificially restore gene flow through the mixing of isolated populations (e.g., Florida panther, *Puma concolor coryi* [25,26]; Gila topminnow, *Poeciliopsis occidentalis* [27]; Greater prairie chicken, *Tympanuchus cupido pinnatus* [28,29]; headwater livebearer, *Poeciliopsis monacha* [30]; Isle Royale wolves, *Canis lupus* [31,32]; mountain pygmy possum, *Burramys parvus* [33]). Such genetic rescue attempts tend to be rare due to biological concerns about outbreeding depression and the potential to lose adaptations to local conditions, as well as cultural concerns about taxonomic integrity and regulatory obstacles [34,35]. However, outcrossing in many cases can be highly beneficial to small, genetically isolated populations that are often inbred, ultimately ensuring their survival [34–37].

Around the time the Chiricahua Leopard Frog Recovery Plan was published in 2007, conservation partners in Arizona had to make difficult choices about a declining, isolated population of Chiricahua leopard frogs on the Coconino National Forest in the Buckskin Hills near Camp Verde, Arizona. Due to a population crash following extensive drought in 2002, the only remaining individuals representing a genetic lineage from the Buckskin Hills population were two males and one female held at the Phoenix Zoo. Conservation partners hoped these frogs would produce offspring for reintroduction, but the pairing did not produce viable egg masses. Attempts were even made to hormonally induce breeding, but with no success.

To retain some of the potentially valuable genetics of this population and have frogs to return to the area, USFWS and AZGFD ultimately decided to cross the remaining Buckskin Hills frogs with individuals from the nearby Gentry Creek lineage located on the Tonto National Forest. Attempts at the Phoenix Zoo to pair two Buckskin Hills males with two Gentry Creek females produced six egg masses in 2008 that had some viable eggs. Though survivorship of larvae was very low, 48 larvae and 18 juveniles and subadults from these crosses were released to the wild in fall of 2008. In subsequent years, additional releases from these crosses helped bolster the reintroduced population, and frogs eventually dispersed across the landscape forming a metapopulation—even moving into areas the local recovery group had not considered releasing them. Within the Buckskin Hills, many sites originally prioritized for recovery have had substantial die-offs from chytridiomycosis and we are now focusing conservation efforts in the dispersal sites where frogs continue to persist.

The decision to cross genetic lineages was relatively straightforward once all other options had been exhausted. However, most Buckskin Hills genetics had already been lost at that point. In hindsight, starting ex situ conservation efforts for the Buckskin Hills lineage earlier may have prevented some of the Buckskin Hills' genetic diversity loss. At the time it did not seem warranted, but the population crashed more quickly than anyone anticipated. Decision making was further hampered because of knowledge gaps in Chiricahua leopard frog genetics. Populations that we eventually crossed were close enough geographically that USFWS, AZGFD, and Phoenix Zoo staff ultimately felt comfortable that the risk of outbreeding depression through a genetic cross was low. At present, Chiricahua leopard frog partners are collaborating with a geneticist who is characterizing the genetic structure and diversity across the species' range to inform an applied genetic management plan for making decisions about timing and need of genetic rescue across populations.

8. Challenges Faced—Example 2: Managing Disease

Disease is another common challenge faced by species, as well as ex situ conservation programs designed to benefit these species. For Chiricahua leopard frogs, as well as many other amphibian species worldwide, chytridiomycosis is a major threat to the survival of populations, hampering recovery efforts [7,38,39]. This skin disease, caused by the fungus *Bd*, leads to mortality in amphibians by disrupting the skin's osmoregulatory function [40]. Sporadic mortality events associated with *Bd* infections have been observed across the range of Chiricahua leopard frogs, especially during the cool season when lower water temperatures in the region are associated with higher prevalence of *Bd* [7,13,14,41]. At present, there is no known preventative measure or treatment for *Bd* that can feasibly be applied in the field for Chiricahua leopard frogs.

Chiricahua leopard frogs of all life stages brought to the Phoenix Zoo for head-starting or breeding are tested for *Bd*, as are water samples from the source site. These incoming individuals are held in a separate quarantine room, where biosecurity protocols [42] are in place to prevent disease transmission to other animals or parts of the facility. Individuals are released from quarantine only after testing negative for *Bd* and ranaviruses, and typically after a quarantine period of 30 days.

At times, the Phoenix Zoo has been called on to take in frogs from areas known to harbor *Bd* due to imminent threats (e.g., severe drought) facing those frog populations. Thus, it has been important to develop an effective strategy for clearing *Bd* in this species ex situ. Multiple treatment options exist for amphibians testing positive for *Bd* in an ex situ setting, including medications and elevated temperature protocols that each have advantages and disadvantages (reviewed in [42]). Different species and/or life stages may vary in their responses to treatments, and multiple treatment types may be needed to clear animals of *Bd* [42]. In 2015, the Phoenix Zoo successfully employed an elevated temperature protocol that cleared quarantined larvae, and larvae that metamorphosed during the quarantine period, of *Bd* [43]. The water temperature in the tanks housing the larvae was elevated for 6 days and larvae were swabbed for *Bd* three times post-treatment. The heat treatment was relatively simple to employ and had no observed short- or long-term adverse effects (e.g., morphological, behavioral, mortality) on treated individuals [43]. Additionally, Zoo staff used this protocol in 2021 with quarantined adult frogs that tested positive for *Bd* to successfully clear them of *Bd* with no adverse effects (Phoenix Zoo, unpublished data). Having an effective *Bd* treatment protocol for use during different life stages allows the Phoenix Zoo and partners to bring in genetic diversity from the wild that we would not be able to access otherwise and is critical for ensuring the health and welfare of all the amphibians in the Zoo's care.

In 2013, the USFWS and AZGFD shifted from requiring prophylactically treating all frogs for *Bd* prior to release with itraconazole to conducting *Bd* testing prior to release to determine if subsequent treatment is necessary. Although effective, prophylactic treatment is costly, time consuming, and stressful to frogs [42].

Amphibian disease, particularly chytridiomycosis, continues to be a challenge to reintroduction efforts, with translocated individuals and populations sometimes succumbing to the disease. Nevertheless, we continue to release Chiricahua leopard frogs into all parts of their historical range in Arizona, including locations with a history of *Bd* [9]. Studies of lowland leopard frogs and Chiricahua leopard frogs found that some populations have

Major Histocompatibility Complex (MHC) alleles associated with increased survival in the presence of *Bd*, along with evidence supporting the hypothesis that *Bd* tolerance is evolving rapidly [44–47]. Additionally, some southern Arizona Chiricahua leopard frog populations appear to co-exist with *Bd* with no apparent population-wide die-offs. Currently we do not fully understand the spatial distribution of *Bd* patterns or seasonal prevalence of *Bd* in the Chiricahua leopard frog's range, or why populations vary in their susceptibility, so additional research is needed to improve the survivorship of Chiricahua leopard frogs in the presence of *Bd*.

We know even less about ranaviruses in Arizona, as widespread testing for the disease has not occurred and relatively little is known about its biology or distribution in Arizona. Two types of ranaviruses, *Amybstoma tigrinum virus* (ATV) and *frog virus* 3 (FV3), are both linked to large episodic die-offs in larval amphibians [48]. While FV3 is not known from Arizona, ATV infects tiger salamanders (*Ambystoma tigrinum*) throughout Arizona but is not known to infect ranids [49,50]. Skin and oral swabs of Chiricahua leopard frogs brought into quarantine at the Phoenix Zoo have revealed that a ranavirus occurs at several wild sites in Arizona. Test results have been difficult to interpret, which is not unusual in captive amphibians that have subclinical infections [42]. Further research is being conducted to understand if and how ranaviruses affect Chiricahua leopard frogs. AZGFD is currently funding research to investigate the prevalence and seasonal dynamics of ranaviruses throughout the Chiricahua leopard frog range and to gain a better understanding of susceptibility of Chiricahua leopard frogs to ranavirus infection.

9. Conclusions

The observed increase in Chiricahua leopard frog site occupancy and breeding since 2007, the presence of functioning metapopulations in several recovery units across Arizona, and the control of bullfrogs in some key areas are significant benchmarks of our progress in Chiricahua leopard frog conservation. Despite multiple partners being actively involved in Chiricahua leopard frog management and progress made toward recovery of the species, long-term persistence of the species in Arizona still requires intensive monitoring and management.

The program described here demonstrates the substantive contributions zoos can make in support of local species conservation. This program is one of numerous amphibian conservation programs with significant zoo and aquarium involvement (as reviewed in [51–55]). Experts emphasize the need for amphibian programs developed and implemented in the animals' native range [52,56–58], which further encourages collaborations like the Chiricahua leopard frog program in Arizona. Significant challenges remain, including disease issues and habitat loss and degradation, but thoughtful, sustained partnerships are essential to the success of local conservation efforts. Adaptive management based on research and monitoring will continue to guide our efforts and hone our methods both in and ex situ.

Supplementary Materials: The following are available online at https://www.mdpi.com/article/10.3390/jzbg3040039/s1, Table S1: Survivorship of *R. chiricahuensis* from hatch to release/transfer, by year and egg mass (EM); Table S2: Number of *R. chiricahuensis* released by year (egg masses excluded).

Author Contributions: Conceptualization, T.R.H., W.L.H., R.A.A., A.K.O., S.H., C.C. and C.A.; writing—original draft preparation, T.R.H., W.L.H., R.A.A. and A.K.O.; writing—review and editing, T.R.H., W.L.H., R.A.A., A.K.O., S.H., C.C. and C.A. All authors have read and agreed to the published version of the manuscript.

Funding: Phoenix Zoo-based conservation efforts detailed here have been funded by the Phoenix Zoo, The Arthur L. and Elaine V. Johnson Foundation, Arizona Game and Fish Department, US Fish and Wildlife Service Arizona Partners for Fish and Wildlife Grant No. 1448-20181-06-G944, National Fish and Wildlife Foundation Sky Island Grasslands Legacy Grant (Project ID: 0103.11.025805), US Forest Service Southern Arizona Resource Advisory Committee Agreement 10-CS-11031200-021, Phoenix Zoo Auxiliary, Association of Zoos and Aquariums Conservation Endowment Fund Grant

2012-1111: Immunogenetics of Chytridiomycosis in Chiricahua Leopard Frogs, and Phoenix Zoo donors with an interest in native species conservation programs. In situ conservation efforts in Arizona are primarily funded by AZGFD, USFWS, and federal land management agencies. The AZGFD Ranid Frogs Project is funded by State Wildlife Grants administered by USFWS, the Arizona Heritage Fund, and Arizona's Nongame Wildlife Checkoff.

Institutional Review Board Statement: Not applicable.

Informed Consent Statement: Not applicable.

Data Availability Statement: The data presented in this study are available in the supplementary material.

Acknowledgments: The authors acknowledge the many individuals whose contributions have been essential to Chiricahua leopard frog recovery and conservation in Arizona, too many to list each by name. We graciously acknowledge Mike Sredl and Jim Rorabaugh for their leadership in establishing and implementing conservation and management of Chiricahua leopard frogs in Arizona. Mike started the Ranid Frogs Project in 1991 and retired from the AZGFD in 2016. Jim Rorabaugh served as the species lead for the USFWS from before it was listed under the Endangered Species Act until he retired in 2011. Mike Demlong initiated the Phoenix Zoo's leopard frog head-starting program in 1995 and dozens of Zoo staff and volunteers have contributed to the success of these efforts from that time on. Special thanks to Tara Sprankle for helping track down historical information included in this manuscript. We also thank other partners involved in the many facets of ex situ and in situ conservation for the Chiricahua leopard frog in Arizona, including Arizona-Sonora Desert Museum, Arizona State University, Bureau of Land Management, Fort Worth Zoo, Ladder Ranch, Marian University, Northern Arizona University, Southwest Research Station, University of Arizona, U.S. Geological Survey, and the numerous private landowners that participate in conservation of the species by having Chiricahua leopard frogs on their properties.

Conflicts of Interest: The authors declare no conflict of interest.

References

1. Allard, R.A.; Wells, S.A. The Phoenix Zoo story: Building a legacy of conservation. In *The Ark and Beyond: The Evolution of Zoo and Aquarium Conservation*; Minteer, B.A., Maienschein, J., Collins, J.P., Eds.; The University of Chicago Press: Chicago, IL, USA, 2018; pp. 169–175.
2. Association of Zoos and Aquariums. *The Accreditation Standards and Related Policies 2022*; Silver Spring: Montgomery County, MD, USA, 2022.
3. U.S. Fish and Wildlife Service. Endangered and threatened wildlife and plants; listing and designation of critical habitat of the Chiricahua leopard frog (*Lithobates chiricahuensis*). *US. Fed. Reg.* **2012**, *77*, 16234–16424.
4. Secretaría de Medio Ambiente y Recursos Naturales. Protección ambiental-especies nativas de México de flora y fauna silvestrescategorías de riesgo y especificaciones para su inclusión, exclusión o cambio-Lista de especies en riesgo. (Segunda Sección). *Diaro Of.* **2010**, *6*, 158–169.
5. Santos-Barrera, G.; Hammerson, G.; Sredl, M. Lithobates chiricahuensis. *IUCN Red List* **2004**, e.T58575A11805575. [CrossRef]
6. U.S. Fish and Wildlife Service. Endangered and threatened wildlife and plants; listing of the Chiricahua leopard frog (*Rana chiricahuensis*). *US. Fed. Reg.* **2002**, *67*, 40790–40811.
7. U.S. Fish and Wildlife Service. *Chiricahua Leopard Frog (Rana chiricahuensis) Recovery Plan*; U.S. Fish and Wildlife Service: Albuquerque, NM, USA, 2007.
8. Rorabaugh, J.C.; Hossack, B.R.; Muths, E.; Sigafus, B.H.; Lemos-Espinal, J.A. Status of the threatened Chiricahua leopard frog and conservation challenges in Sonora, Mexico, with notes on other ranid frogs and non-native predators. *Herpetol. Conserv. Biol.* **2018**, *13*, 17–32.
9. Mosley, C.D.; Marsh, M.J.L.; Owens, A.K. Chiricahua leopard frog recovery in Arizona 2019. In *Nongame and Endangered Wildlife Program Technical Report 330*; Arizona Game and Fish Department: Phoenix, AZ, USA, 2020.
10. Hossack, B.R.; Howell, P.E.; Owens, A.K.; Cobos, C.; Goldberg, C.S.; Hall, D.; Hedwall, S.; MacVean, S.K.; MacCaffery, M.; McCall, A.H.; et al. Identifying factors linked with persistence of reintroduced populations: Lessons learned from 25 years of amphibian translocations. *Global Ecol. Conserv.* **2022**, *35*, e02078. [CrossRef]
11. Demlong, M.J. Head-starting *Rana subaquavocalis* in captivity. *Reptiles* **1997**, *5*, 24–33.
12. Sprankle, T. Giving leopard frogs a head start. *Endanger. Species Update* **2008**, *25*, S15+.
13. Sredl, M.J.; Field, K.J.; Peterson, A.M. Understanding and mitigating effects of chytrid fungus to amphibian populations in Arizona. In *Nongame and Endangered Wildlife Program Technical Report 208*; Arizona Game and Fish Department: Phoenix, AZ, USA, 2003.

14. Sredl, M.J.; Jennings, R.D. *Rana chiricahuensis* Platz and Mecham, 1979: Chiricahua leopard frogs. In *Amphibian Declines: The Conservation Status of United States Species*; Lanoo, M., Ed.; University of California Press: Berkeley, CA, USA; Los Angeles, CA, USA, 2005; pp. 546–549.
15. Vitt, L.J.; Caldwell, J.P. *Herpetology: An Introductory Biology of Amphibians and Reptiles*, 4th ed.; Academic Press: San Diego, CA, USA, 2014.
16. Reading, R.P.; Miller, B.J. The black-footed ferret recovery program: Unmasking professional and organizational weaknesses. In *Endangered Species Recovery: Finding the Lessons, Improving the Process*; Clark, T.W., Reading, R.P., Clarke, A.L., Eds.; Island Press: Washington, DC, USA, 1994; pp. 73–100.
17. Scott, J.M.; Goble, D.D.; Wiens, J.A.; Wilcove, D.S.; Bean, M.; Male, T. Recovery of imperiled species under the Endangered Species Act: The need for a new approach. *Front. Ecol. Environ.* **2005**, *3*, 383–389. [CrossRef]
18. Dreitz, V. Issues in species recovery: An example based on the Wyoming toad. *Bioscience* **2006**, *56*, 765–771. [CrossRef]
19. Miller, B.; Reading, R.P. Challenges to black-footed ferret recovery: Protecting prairie dogs. *West. N. Am. Nat.* **2012**, *72*, 228–240. [CrossRef]
20. Evans, D.M.; Che-Castaldo, J.P.; Crouse, D.; Davis, F.W.; Epanchin-Niell, R.; Flather, C.H.; Frohlich, R.K.; Goble, D.D.; Li, Y.; Male, T.D.; et al. Species recovery in the United States: Increasing the effectiveness of the Endangered Species Act. *Issues Ecol.* **2016**, *20*, 1–28.
21. Walls, S.C.; Ball, L.C.; Barichivich, W.J.; Dodd, C.K., Jr.; Enge, K.M.; Gorman, T.A.; O'Donnell, K.M.; Palis, J.G.; Semlitsch, R.D. Overcoming challenges to the recovery of declining amphibian populations in the United States. *Bioscience* **2017**, *67*, 156–165. [CrossRef]
22. Rorabaugh, J.; Kreutzian, K.; Sredl, M.; Painter, C.; Aguilar, R.; Bravo, J.C.; Kruse, C. Chiricahua leopard frog inches toward recovery. *Endanger. Species Update* **2008**, *25*, S10+.
23. Byers, O.; Lees, C.; Wilcken, J.; Schwitzer, C. The One Plan Approach: The philosophy and implementation of CBSG's approach to integrated species conservation planning. *WAZA Mag.* **2013**, *14*, 2–5.
24. Traylor-Holzer, K.; Leus, K.; Byers, O. Integrating ex situ management options as part of a One Plan Approach to species conservation. In *The Ark and Beyond: The Evolution of Zoo and Aquarium Conservation*; Minteer, B.A., Maienschein, J., Collins, J.P., Eds.; University of Chicago Press: Chicago, IL, USA, 2018; pp. 129–141.
25. Hedrick, P.W. Gene flow and genetic restoration: The Florida panther as a case study. *Conserv. Biol.* **1995**, *9*, 996–1007. [CrossRef]
26. Pimm, S.L.; Dollar, L.; Bass, O.L., Jr. The genetic rescue of the Florida panther. *Anim. Conserv.* **2006**, *9*, 115–122. [CrossRef]
27. Hedrick, P.W.; Lee, R.; Hurt, C.R. Genetic evaluation of captive populations of endangered species and merging of populations: Gila topminnows as an example. *J. Hered.* **2012**, *103*, 651–660. [CrossRef]
28. Westemeier, R.L.; Brawn, J.D.; Simpson, S.A.; Esker, T.L.; Jansen, R.W.; Walk, J.W.; Kershner, E.L.; Bouzat, J.L.; Paige, K.N. Tracking the long-term decline and recovery of an isolated population. *Science* **1998**, *282*, 1695–1698. [CrossRef]
29. Warnke, K. Wisconsin greater prairie-chicken management plan 2004–2014. In *Wisconsin Department of Natural Resources*; Madison: Wisconsin, WI, USA, 2004.
30. Vrijenhoek, R.C. Genetic diversity and fitness in small populations. In *Conservation Genetics EXS*; Loeschcke, V., Jain, S.K., Tomiuk, J., Eds.; Birkhäuser: Basel, Switzerland, 1994; Volume 68.
31. Räikkönen, J.; Vucetich, J.A.; Peterson, R.O.; Nelson, M.P. Congenital bone deformities and the inbred wolves (*Canis lupus*) of Isle Royale. *Biol. Conserv.* **2009**, *142*, 1025–1031. [CrossRef]
32. Robinson, J.A.; Räikkönen, J.; Vucetich, L.M.; Vucetich, J.A.; Peterson, R.O.; Lohmueller, K.E.; Wayne, R.K. Genomic signatures of extensive inbreeding in Isle Royale wolves, a population on the threshold of extinction. *Sci. Adv.* **2019**, *5*, eaau0757. [CrossRef] [PubMed]
33. Weeks, A.R.; Moro, D.; Thavornkanlapachai, R.; Taylor, H.R.; White, N.E.; Weiser, E.L.; Heinze, D. Conserving and enhancing genetic diversity in translocation programs. In *Advances in Reintroduction Biology of Australian and New Zealand Fauna*; Armstrong, D., Hayward, M., Moro, D., Seddon, P., Eds.; CSIRO Publishing: Melbourne, Australia, 2015.
34. Frankham, R. Genetic rescue of small inbred populations: Meta-analysis reveals large and consistent benefits of gene flow. *Molec. Ecol.* **2015**, *24*, 2610–2618. [CrossRef] [PubMed]
35. Ralls, K.; Ballou, J.D.; Dudash, M.R.; Eldridge, M.D.B.; Fenster, C.B.; Lacy, R.C.; Sunnucks, P.; Frankham, R. Call for a paradigm shift in the genetic management of fragmented populations. *Conserv. Lett.* **2018**, *11*, 1–6. [CrossRef]
36. Whiteley, A.R.; Fitzpatrick, S.W.; Funk, W.C.; Tallmon, D.A. Genetic rescue to the rescue. *Trends Ecol. Evol.* **2015**, *30*, 42–49. [CrossRef]
37. Frankham, R. Genetic rescue benefits persist to at least the F3 generation, based on a meta-analysis. *Biol. Conserv.* **2016**, *195*, 33–36. [CrossRef]
38. Longcore, J.E.; Pessier, A.P.; Nichols, D.K. *Batrachochytrium dendrobatidis* gen. et sp. nov., a chytrid pathogenic to amphibians. *Mycologia* **1999**, *91*, 219–227. [CrossRef]
39. Skerratt, L.F.; Berger, L.; Speare, R.; Cashins, S.; McDonald, K.R.; Phillott, A.D.; Hines, H.B.; Kenyon, N. Spread of chytridiomycosis has caused the rapid global decline and extinction of frogs. *EcoHealth* **2007**, *4*, 125. [CrossRef]
40. Voyles, J.; Young, S.; Berger, L.; Campbell, C.; Voyles, W.F.; Dinudom, A.; Cook, D.; Webb, R.; Alford, R.A.; Skerratt, L.F.; et al. Pathogenesis of chytridiomycosis, a cause of catastrophic amphibian declines. *Science* **2009**, *326*, 582–585. [CrossRef]

41. Forrest, M.J.; Schlaepfer, M.A. Nothing a hot bath won't cure: Infection rates of amphibian chytrid fungus correlate negatively with water temperature under natural field settings. *PLoS ONE* **2011**, *6*, e28444. [CrossRef]
42. Pessier, A.P.; Mendelson, J.R., III. *A Manual for Control of Infectious Diseases in Amphibian Survival Assurance Colonies and Reintroduction Programs*; Version 2.0; IUCN/SSC Conservation Breeding Specialist Group: Apple Valley, MN, USA, 2017.
43. Heuring, W.L.; Poynter, B.M.; Wells, S.; Pessier, A.P. Successful clearance of chytrid fungal infection in threatened Chiricahua leopard frog (*Rana chiricahuensis*) larvae and frogs using an elevated temperature treatment protocol. *Salamandra* **2021**, *57*, 171–173.
44. Savage, A.E.; Zamudio, K.R. MHC genotypes associate with resistance to a frog-killing fungus. *Proc. Natl. Acad. Sci. USA* **2011**, *108*, 16705–16710. [CrossRef] [PubMed]
45. Savage, A.E.; Becker, C.G.; Zamudio, K.R. Linking genetic and environmental factors in amphibian disease risk. *Evol. Appl.* **2015**, *8*, 560–572. [CrossRef]
46. Savage, A.E.; Zamudio, K.R. Adaptive tolerance to a pathogenic fungus drives major histocompatibility complex evolution in natural amphibian populations. *Proc. R. Soc. B* **2016**, *283*, 20153115. [CrossRef] [PubMed]
47. Savage, A.E.; Mulder, K.P.; Torres, T.; Wells, S. Lost but not forgotten: MHC genotypes predict overwinter survival despite depauperate MHC diversity in a declining frog. *Conserv. Genet.* **2018**, *19*, 309–322. [CrossRef]
48. Miller, D.; Gray, M.; Storfer, A. Ecopathology of ranaviruses infecting amphibians. *Viruses* **2011**, *3*, 2351–2373. [CrossRef]
49. Jancovich, J.K.; Davidson, E.W.; Morado, J.F.; Jacobs, B.L.; Collins, J.P. Isolation of a lethal virus from the endangered tiger salamander *Ambystoma tigrinum stebbinsi*. *Dis. Aquat. Organ.* **1997**, *31*, 161–167. [CrossRef]
50. Jancovich, J.K.; Davidson, E.W.; Seiler, A.; Jacobs, B.L.; Collins, J.P. Transmission of the *Ambystoma tigrinum* virus to alternative hosts. *Dis. Aquat. Organ.* **2001**, *46*, 159–163. [CrossRef]
51. Browne, R.K.; Wolfram, K.; García, G.; Bagaturov, M.F.; Pereboom, Z.J.J.M. Zoo-based amphibian research and conservation breeding programs. *Amphib. Reptile Conserv.* **2011**, *5*, 1–14.
52. Zippel, K.; Johnson, K.; Gagliardo, R.; Gibson, R.; McFadden, M.; Browne, R.; Martinez, C.; Townsend, E. The Amphibian Ark: A global community for *ex situ* conservation of amphibians. *Herpetol. Conserv. Biol.* **2011**, *6*, 340–352.
53. Harding, G.; Griffiths, R.A.; Pavajeau, L. Developments in amphibian captive breeding and reintroduction programs. *Conserv. Biol.* **2016**, *30*, 340–349. [CrossRef]
54. Murphy, J.B.; Gratwicke, B. History of captive management and conservation amphibian programs mostly in zoos and aquariums. *Herp. Rev.* **2017**, *48*, 241–260.
55. Mendelson, J.R., III. Frogs in glass boxes: Responses of zoos to global amphibian extinctions. In *The Ark and Beyond: The Evolution of Zoo and Aquarium Conservation*; Minteer, B.A., Maienschein, J., Collins, J.P., Eds.; University of Chicago Press: Chicago, IL, USA, 2018; pp. 298–310.
56. Gagliardo, R.; Crump, P.S.; Griffith, E.J.; Mendelson, J.; Ross, H.; Zippel, K. The principles of rapid response for amphibian conservation, using the programmes in Panama as an example. *Int. Zoo Yearb.* **2008**, *42*, 125–135. [CrossRef]
57. Mendelson, J.R., III; Lips, K.R.; Gagliardo, R.W.; Rabb, G.B.; Collins, J.P.; Diffendorfer, J.E.; Daszak, P.; Ibáñez, D.; Zippel, K.C.; Lawson, D.P.; et al. Confronting amphibian declines and extinctions. *Science* **2006**, *313*, 48. [CrossRef] [PubMed]
58. Carrillo, L.; Johnson, K.; Mendelson, J., III. Principles of program development and management for amphibian conservation captive breeding programs. *Int. Zoo News* **2015**, *62*, 96–107.

Journal of *Zoological and Botanical Gardens*

MDPI

Article

Collaborative Conservation by Botanical Gardens: Unique Opportunities for Local to Global Impacts

Aireona B. Raschke [1,2,*], Kimberly V. Pegram [1], Natalie A. Melkonoff [1], Jeny Davis [1] and Steven A. Blackwell [1]

1 Desert Botanical Garden, Research, Conservation and Collections, Phoenix, AZ 85008, USA
2 Center for Collaborative Conservation, Colorado State University, Fort Collins, CO 80523, USA
* Correspondence: araschke@colostate.edu

Abstract: Conservation organizations with common missions can find strength and synergy in collaboration. Collaboration can also be challenging, especially finding the right partnerships or organizations to lead. Within the "ecosystem" of conservation organizations, botanical gardens have a unique array of resources and expertise which make them ideal candidates for leadership or partnership in collaborative conservation efforts. We will explore this idea by examining four conservation initiatives at Desert Botanical Garden (Phoenix, AZ, USA) that approach collaborative conservation on regional, state, and international scales. On a regional scale, Metro Phoenix EcoFlora and the Central Arizona Conservation Alliance lead landscape-level conservation by providing a structure for more than 60 official conservation partners, by generating data, and through public engagement needed in a rapidly developing region. On the state scale, Great Milkweed Grow Out is an initiative for pollinator conservation that provides expertise, materials, and opportunities for a wide range of partners across Arizona. Desert Botanical Garden's endangered plant species conservation efforts provide expertise and resources through horticulture and seed preservation for threatened and endangered plants across the US and internationally. We will share the structure of each program where applicable, how they came to fruition, and their successes. Through each case study, we will highlight the ways positioning within a botanical garden has benefitted the program and success in collaboration. We will also highlight unique challenges. Botanical gardens provide unique opportunities, and they should not be overlooked when seeking a conservation partner or leader.

Keywords: botanical gardens; conservation; collaboration; natural resource management; outreach; science communication; endangered species; pollinator; conservation horticulture; seed banking

Citation: Raschke, A.B.; Pegram, K.V.; Melkonoff, N.A.; Davis, J.; Blackwell, S.A. Collaborative Conservation by Botanical Gardens: Unique Opportunities for Local to Global Impacts. *J. Zool. Bot. Gard.* **2022**, 3, 463–487. https://doi.org/10.3390/jzbg3030035

Academic Editors: Ursula Bechert and Debra C. Colodner

Received: 3 August 2022
Accepted: 12 September 2022
Published: 19 September 2022

1. Introduction

Human impacts worldwide have generated the urgent need for conservation of natural resources. In modern times, this work has involved organizations of all kinds, including government agencies, academic institutions, NGOs and non-profits, industry, land managers/owners, and communities. While progress has been made to protect the environment through the individual efforts of these entities, the increasing rapidity and complexity of biodiversity loss has made it clear that collaborative conservation efforts are essential to effectively address this global threat. Collaboration, however, is a complex process that requires skilled and dedicated leaders for its successful implementation. Botanical gardens are uniquely suited as backbone organizers and leaders through the expertise, resources, and common ground that they represent. This paper will explore the role of botanical gardens in collaborative conservation through the close examination of several Desert Botanical Garden (Phoenix, AZ, USA) case studies.

1.1. Collaborative Conservation

Collaborative conservation includes efforts to preserve, protect, and/or sustainably manage natural resources by two or more entities explicitly working together, often including shared goal setting, decision making, and implementation [1]. Joint efforts to manage natural resources equitably and sustainably go back to ancient times, but modern, Western conservation efforts have most often emphasized the centralized decision making of single organizations and top-down management until recently [2,3]. Increasingly, collaborative efforts are now acknowledged [4,5] as being essential to conservation due to their potential to address landscape-scale conservation needs that transcend management boundaries, include common pool resources and/or limited resources for management (e.g., funding and staff capacity), and the pressing need for diverse perspectives and knowledge types to support innovation [1,6,7].

A robust knowledge base has been developed that examines the complex process that collaborative conservation efforts entail and deepens our understanding of the factors that lead to success. In particular, two bodies of knowledge apply to the work represented here. The first is Ostrom's Design Principles [8], which define factors for success among efforts to jointly manage common-pool resources. These include defined boundaries; governance with fit-to-place, participatory decision making and the right to organize; monitoring of resources; graduated sanctions; access to conflict resolution; and the presence of nested networks [9]. While we will not be delving into this framework in this paper, several of the case studies presented concern common-pool resources and implicitly include consideration of the design principles in their structure. More explicit is the integration of the collective impact framework into the efforts of the Central Arizona Conservation Alliance (CAZCA), which is considered among the case studies. Collective impact focuses on the structures necessary for effective collaborative effort including (1) backbone staff to facilitate the collaborative efforts, (2) a common agenda among partners, (3) shared measurement of results, (4) mutually reinforcing activities throughout the collaboration, and (5) open and continuous communication [10].

1.2. Botanical Gardens and Conservation

Botanical gardens and arboreta (henceforth jointly referred to as botanical gardens) around the world display living plants and provide opportunities for visitors to learn about and enjoy them. They also contribute to plant conservation on all scales—local to global. In most cases, these living plants serve as collections for ex situ conservation. It is estimated that 25–30% of plant species are accessioned in living collections of botanical gardens globally [11–14]. Seed banks house collections that exceed 55,000 plant species from around the world, which is valuable for the preservation of genetic diversity [15,16]. Botanical garden herbaria provide depositories of plant specimens for documentation and research totaling more than 140 million specimens worldwide [17]. They contain valuable data on plant diversity, distribution, and historical records that can serve as a baseline for studies of climate change and species assessments. Scientists and staff at botanical gardens and collaborating institutions use living collections, seed banks, and herbaria to study all levels of botany, including taxonomy, plant biology, physiology, ecology, and conservation [17,18]. Scientists serving botanical gardens often also participate in in situ conservation [19,20].

Beyond research and collections, botanical gardens can support or lead conservation efforts through public education [14,21,22]. Programs, visitor centers, and classes can teach learners of all ages about plants, ecosystems, and climate change while countering "plant invisibility" [21–24]. They can also engage people in the appreciation of plants through community science initiatives that encourage participants to seek out flora while providing crucial data to scientists [14,25,26].

Botanical gardens have key resources and qualities that enhance their ability to serve as centers for collaborative conservation. This paper will highlight some of these benefits and give examples through case studies of four programs or initiatives at Desert Botanical Garden (DBG). Each case study will describe the development of the collaboration, key successes and quantifiable results, and how positioning within DBG has fostered these successes. The discussion will explore some of the challenges faced by collaborative botanical garden conservation efforts.

2. Materials and Methods

The focal institution examined in this paper is the Desert Botanical Garden (DBG), located in Phoenix, AZ, USA. As of 2022, DBG manages 140 acres of Sonoran Desert habitat with over 50,000 arid ecosystem plants. These plants consist of more than 4800 taxa, over 500 of which are of conservation concern [27]. DBG has a growing base of skilled staff, hundreds of volunteers, and welcomes an average of 450,000 visitors a year.

DBG was founded in 1939 to conserve desert plants and habitat in a rapidly expanding city and has maintained this commitment to conservation. In 1985 it became a charter member of the Center for Plant Conservation, in 2012 it was a founding member of the Central Arizona Conservation Alliance (CAZCA), and in 2017 DBG completed the Hazel Hare Center for Plant Science (HHCPS). Its most recent strategic plan also carries the mission forward with an entire section focused on environmental sustainability and plant conservation [28].

The focus of this paper will be on the collaborative initiatives and projects of DBG that demonstrate the role of botanical gardens in local, regional, national, and international collaborative conservation efforts. These include the Metro Phoenix EcoFlora (MPE), CAZCA, Great Milkweed Grow Out (GMGO), and DBG's endangered plant conservation program and efforts (Table 1).

Table 1. Summary of DBG case study programs/projects.

Program/ Project Name	Conservation Focus	Scale	No. of Collaborators	Types of Partner Organizations
EcoFlora	Urban biodiversity, alleviating plant invisibility, open-access data and information	Local—Metro Phoenix area	40+	Non-profits, academic, municipal agencies, societies, professional scientists, local experts, community members
CAZCA	Habitat connectivity, restoration, outreach, collaborative coordination	Regional—Maricopa County and associated HUC watersheds	70+	Federal, state and municipal agencies, non-profits, academic institutions, industry
GMGO	Pollinators, native plant materials, research, outreach	State	60+	Federal and state agencies, non-profits, academic institutions, community members, K-12 schools
Endangered Species Plant Conservation	Rare, threatened, and endangered plant conservation ex situ and in situ	International	70+	Federal and state agencies, IUCN, other botanical research institutions

2.1. EcoFlora

EcoFlora is an initiative of the New York Botanical Garden (NYBG) created in 2016. The program is focused on understanding urban biodiversity, increasing access to biodiversity data and information, and alleviating plant invisibility. Simultaneously, the project bolsters herbarium collections, documents plants and their associated organisms, and engages with the public. Observations are recorded by community scientists using iNaturalist (https://www.inaturalist.org/ accessed on 15 February 2022), a free and open-source, web-based platform and app that is a joint initiative by the California Academy of Sciences

(San Francisco, CA, USA) and the National Geographic Society (Washington, DC, USA). iNaturalist engages users by functioning as a digital field guide, suggesting identifications and providing taxon information, while also serving as a database that includes images and geospatial data. EcoFlora engages the public through presentations, training, events, educational resources, and communications.

In 2019, a National Leadership Grant from the Institute of Museum and Library Services was awarded to NYBG to expand the program within the United States, and four partner botanical gardens were selected to implement the EcoFlora model in their respective locations. DBG was selected, along with Chicago Botanic Garden, Denver Botanic Gardens, and Marie Selby Botanical Gardens. The projects are collectively known as the EcoFloras of North America. Each garden operates their EcoFlora project autonomously, while collaborating under the guidance of NYBG.

The EcoFlora program at DBG is known as the Metro Phoenix EcoFlora (MPE). It began in February 2020 and operates within the CAZCA initiative (Section 2.2). The study area includes metro Phoenix (AZ, USA) and is estimated to cover 2700 square miles (4300 square kilometers) (Figure 1). MPE prioritizes community engagement and collaboration, emphasizing that awareness is the first step in increasing appreciation of plant life and environmental literacy, thereby fostering public involvement in conservation [29]. The project collaborates with scientists, organizations, and community members in events, communications, resources, and data collection. Additionally, the iNaturalist observations and information gathered in the metro Phoenix area complement existing floristic data in the Southwest Environmental Information Network (SEINet; https://swbiodiversity.org/seinet/ accessed on 15 February 2022) and the DBG herbarium collections.

Figure 1. Study area for the Metro Phoenix EcoFlora project created with iNaturalist, using Google Maps data.

2.2. Central Arizona Conservation Alliance

CAZCA is an initiative of DBG designed to lead regional conservation collaboration through strategic planning and convenings, as well as programmatic work. The initiative was founded in 2012, with leadership from DBG along with Arizona State University, Maricopa County Parks and Recreation Department, and others. The motivation for this effort was the rapid population growth in Maricopa County and associated biodiversity loss due to development practices [30,31]. It was believed that collaborative leveraging of limited resources and sharing of expertise through CAZCA would support the increased flexibility, adaptability, and scale of conservation efforts necessary to protect ecosystem health while balancing the needs of a growing population. CAZCA's primary spatial focus is on Maricopa County in Arizona, USA, with additional areas for associated Hydraulic Unit Code (HUC) 10 watersheds (Figure 2) to increase the ecological relevance of CAZCA's work [32].

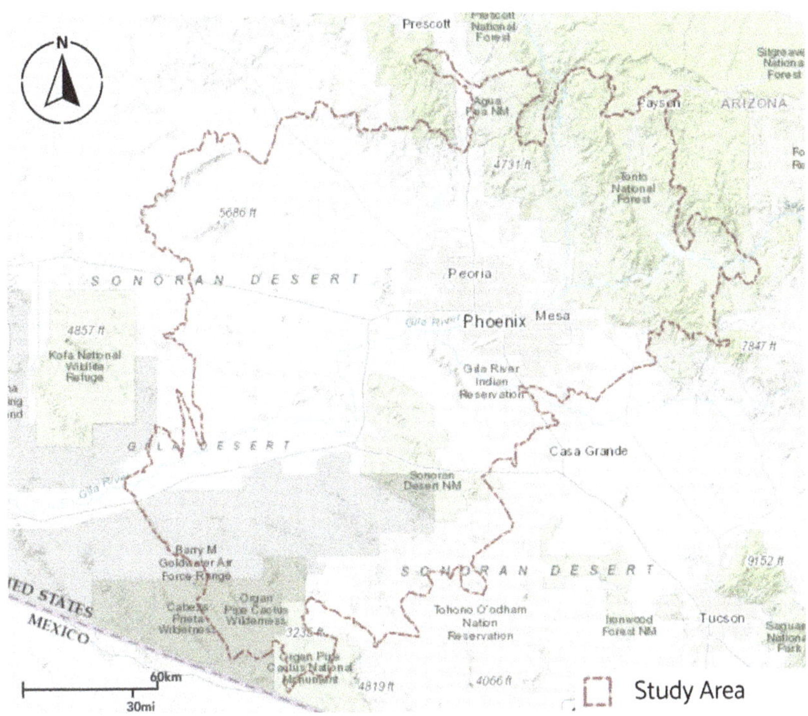

Figure 2. Map of the CAZCA study area in Arizona, USA.

Structurally, CAZCA follows the Collective Impact model [10], with DBG serving as the backbone organization that houses the staff dedicated to the collaborative work of the initiative. At the time of writing this article (2022), CAZCA had a staff of two, a program director and an engagement coordinator, three dedicated volunteers, and annual contracted expertise. The initiative is further supported and advised by a steering committee with representation from six organizations from among the 70+ partner organizations that make up the Alliance.

The work of the initiative is guided by the Regional Open Space Strategy for Maricopa County (ROSS). The ROSS was the first major CAZCA project and was developed through 18 months of iterative discussion among 73 participating organizations (Appendix A). The focus of these convenings was to develop goals and objectives that would be necessary to attain the shared vision of a regional network of natural open spaces across Maricopa

County. This network of natural open spaces, as envisioned, would be structured and managed to support biodiversity and human well-being. This resulted in a finalized document published in 2018, with four primary goals outlined by the partner organizations [33]:

1. Protect and Connect—Ensure a robust network of habitat blocks and connections to sustain native plant and animal communities, provide opportunities for recreation, support clean air and water resources, and improve resilience to drought, extreme heat, and flooding.
2. Sustain and Restore—Identify and engage best practices in land management and restoration to sustain and enhance native biodiversity, positive recreational experiences, and socio-economic benefits connected with the Sonoran Desert.
3. Love and Support—Build champions and constituency of support and action for Sonoran Desert conservation by raising awareness and connecting people with nature.
4. Coordinate and Elevate—Build upon the CAZCA foundation to ensure and amplify regional open space collaboration, coordination, management, and conservation successes.

Since 2018, CAZCA has shifted its focus to accomplishing the objectives outlined by the ROSS. This work is carried out through two methods: (1) facilitating and supporting continued and targeted convenings; and (2) innovative, collaborative project incubation. In both cases, the initiative staff consult with partner organizations on the identification of priority activities, groups are formed to plan for activities from among the partners, and projects are carried out with shared capacity.

2.3. *Great Milkweed Grow Out*

GMGO is an initiative of DBG started in 2016 to support the conservation of monarch butterflies and other pollinators. Monarch butterflies have faced drastic declines in their two US migratory flyways over the last three decades [34–36]. While there are likely multiple causes of decline, most authorities agree that habitat loss is a major contributor and that habitat restoration is a key conservation tool [35,37]. Of particular concern is the loss of monarch breeding habitat, which includes milkweed plants (*Asclepias* spp.) and nectar-producing plants as its primary components [38–40]. Monarchs can benefit from small habitats in urban areas [41,42], so urban plantings provide a direct method through which individuals can help protect a beloved butterfly species.

Native plants are essential to monarch survival and, with expertise in native plants, botanical gardens can serve as leaders in monarch conservation. At DBG, monarch conservation messages have been shared with visitors for many years through exhibitions and volunteer interactions, largely through the Jonathan and Maxine Marshall Butterfly Exhibit and the newer Butterfly Exhibit opened in 2017. Visitors are encouraged to create habitat for monarchs with the use of native plants, but historically this call to action was hard to execute. Native milkweeds were not often available in plant nurseries and there was a general lack of understanding of how monarchs interacted with milkweeds in the region [43,44]. This need for plants and a greater understanding of butterfly–plant interactions led to the development of GMGO as an internal collaboration between the Exhibits Department, with their butterfly expertise in the Butterfly Pavilion, and the Research, Conservation, and Collections Department, with research and native plant propagation expertise. GMGO has since expanded to include three full-time staff with seasonal interns who collaborate with a variety of partners statewide.

The structure of GMGO was developed to combine the basic needs of monarch conservation in Arizona with the strengths of DBG. It has three major activities that are all intertwined and integral to success: plant propagation, ecological research, and community outreach. Propagation of native milkweeds provides the basic resources needed for habitat plantings, outreach, and researching monarch–milkweed interactions. GMGO focuses propagation efforts on plants that are not common in commercial wholesale nurseries and thus often require some experimentation with seed germination and growing protocols to efficiently produce viable plants from wild-collected seed. Scientific research into monarch–

milkweed interactions fills essential knowledge gaps and provides the information needed to use limited resources most effectively. Outreach facilitates the distribution of milkweed plants throughout the Phoenix area while sharing the importance of native plants with the public. Combining these activities into one conservation initiative allows for enhanced conservation outcomes via the integration of key activities and the joint leveraging of partner resources and expertise, leading to more engagement and conservation action, as well as more chances for collaboration.

2.4. Rare, Threatened, and Endangered Species Conservation

The threats to plants and their associated ecosystems are numerous and ever-increasing. In the US and Canada there have been an estimated 65 vascular plant taxa lost to extinction since European colonization [45]. Focused on plants found in the Southwest, cacti, and succulents, DBG is conserving plants to avoid future extinctions. Threatened and endangered plant species are maintained at DBG in long-term storage in the seed bank and as part of the living collection, both on public display and in cultivation in the state-of-the-art facilities within the HHCPS.

In 2019, the Ahearn Desert Conservation Laboratory (DCL) opened as part of the HHCPS. Key features of this new facility are a seed bank and seed lab that houses growth chambers used for testing seed viability, developing germination protocols, and in vitro propagation through tissue culture. In addition to the long-term preservation of rare species, the seed collection is also used to enhance the living collection for research, seed amplification, and in situ restoration efforts. DBG also employs personnel with extensive experience in collection, horticulture, and seed germination research.

Researchers realized early in the history of the Garden that to effectively manage the ever-growing list of threatened species, collaboration would have to be a key strategy of DBG's plant conservation efforts. DBG is a charter member of the Center for Plant Conservation (CPC), which coordinates a network of 71 botanical institutions that work collectively to conserve over 2200 of the most imperiled species of North America, referred to as the "CPC National Collection of Endangered Plants" [46]. DBG currently manages 62 taxa within the national collection that spans seven states and ranges geographically from California to Florida. Under a similar agreement, in 2018, DBG signed an MOU with the North American Orchid Conservation Center, founded by the Smithsonian Institute and U.S. Botanic Garden [47]. This agreement gave the Garden the distinction of becoming the regional seed bank for the more than 30 orchid taxa that inhabit the Southwestern region [48]. As the designated seed bank, the aim is to work with other regional partners to collect and preserve every taxon throughout the Southwest.

DBG also contributes to rare plant conservation by supporting government agencies that are mandated to manage sensitive species on public lands. Since over 75% of lands in Arizona are estimated to be managed by the federal government, Native American tribes, or the state [49], this can be an overwhelming challenge. Adding to this challenge, agencies often lack sufficient funding and, as a result, staff dedicated to managing rare plants on these lands. For these reasons, they often rely on the expertise and resources available at botanical gardens and other botanical institutions to help manage rare species that fall under their jurisdiction. This is illustrated by two projects: one to salvage the endangered Arizona hedgehog cactus (*Echinocereus arizonicus*) impacted by the construction of a new bridge and another to survey and propagate an endangered orchid, the Canelo Hills ladies' tresses (*Spiranthes delitescens*), on the brink of extinction.

Another collaboration in the conservation of plant species is DBG's participation in the International Union for the Conservation of Nature (IUCN). In 2015, DBG became the host institution for the IUCN Cactus and Succulent Specialist Group (CSSG) as part of the Species Survival Commission. As of 2021, DBG employs a Red List Authority Coordinator that oversees the completion and publishing of Red List species assessments and the CSSG Programme Officer. DBG also contributes staff expertise to this group to assess and guide cactus and succulent conservation efforts worldwide.

group met more than 40 times during this process. Since its completion, the ROSS has led to regional planning and collaboration outcomes, including: (1) the development of tools from shared resources; (2) increased collaboration around large-scale challenges, such as habitat connectivity and land acquisition; and (3) the successful launching of regional conservation efforts, such as landscape-scale mapping and management of invasive plant species.

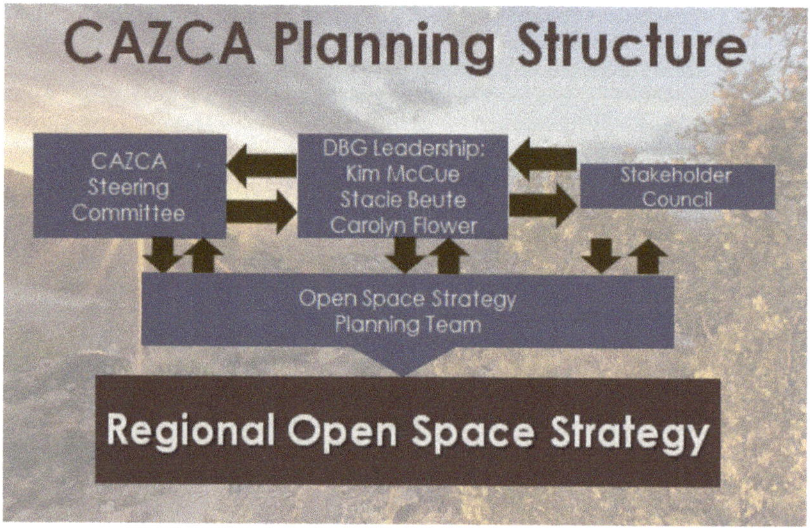

Figure 4. Phase one of the structure that CAZCA utilized in planning the ROSS.

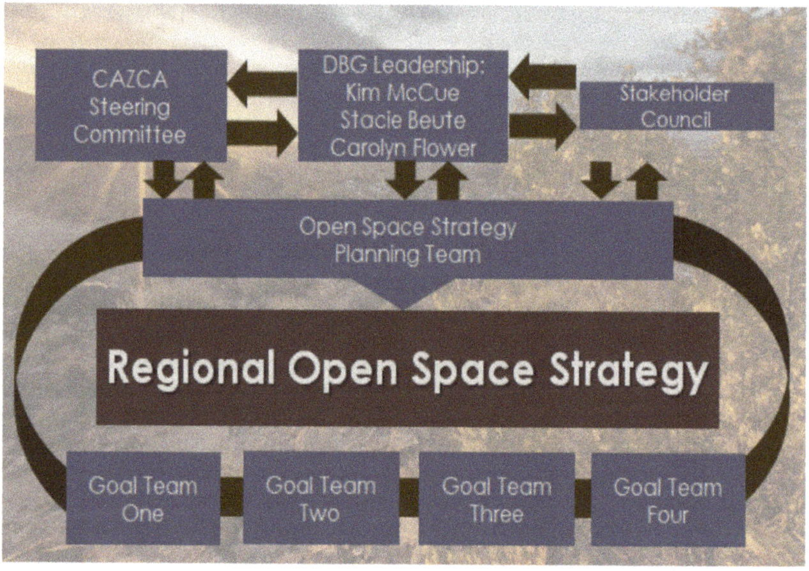

Figure 5. Phase two of the structure that CAZCA utilized in planning the ROSS.

CAZCA has also provided facilitation and organizational support to partner organizations, as well as a neutral space for professionals to collaborate on the creation of tools that will benefit the region. In many cases, without the backbone support of CAZCA, these products would not have been possible.

The first of the tools developed is a natural infrastructure viewer called the Greenprint (accessible at https://web.tplgis.org/cazca_plan/ accessed on 1 February 2022); this includes data on natural resources across Maricopa County, from water to habitat quality, as well as data relevant to urban planning and habitat preservation, such as parcel-level data and heat vulnerability information, among many other elements. This is called for in the ROSS, in Objective 1.1, and its development required in-depth partner collaboration. A team of technical experts was assembled by CAZCA to determine what data to include and to vet the tool in a cyclical development process that was led by the Trust for Public Land. The Greenprint was successfully unveiled to the partner organizations in 2019 and has been utilized by a variety of regional planning and grassroots efforts since then, including Maricopa County's Parks 2030 plan.

CAZCA was also the facilitating and convening force in the creation of a collaborative plan for habitat connectivity in Central Arizona. The process of developing the plan involved iterative gatherings of an advisory team and a larger regional stakeholder meeting which involved 35 partner organizations (see Appendix B for a full list). The focus of these meetings was on synthesizing spatial data and expert knowledge in order to identify what were termed Conservation Opportunity Areas (COAs). COAs were large, natural areas identified by the stakeholders as important to conservation and practical for preservation as rapid urban development continues in the region.

In conjunction with the planning and tool development described above, CAZCA has been instrumental in the creation of a more collaborative environment in Maricopa County for efforts related to biodiversity conservation and nature-based community well-being. This includes CAZCA's direct collaborative project work and the general increase in collaborations across the region post-ROSS development. First, the Sonoran Seed Collaborative project involves eight partner organizations in the development and study of genetically appropriate native plant materials for habitat restoration. To date, it has produced two annual reports tracking the scientific outcomes of the project, it has funded and carried out habitat restoration at Papago Park and Piestewa Peak in Phoenix, AZ, and the group hosted a 2022 Sonoran Seed Summit that supported expert exploration of how to increase production of native plant materials in Central Arizona.

CAZCA also runs a collaborative project called Desert Defenders, which involves seven partner organizations and focuses on the management of nine invasive plant species. The project trains community scientists to map invasive plants on trails with their smartphones, has created a public map of this spatial data, and serves as a platform for sharing knowledge on invasive plant management strategies among partners. From 2018 to 2022, it has mapped hundreds of acres and identified 6532 invasive plant occurrences.

Finally, when CAZCA was founded in 2012, it was with the realization among partner organizations that collaboration was lacking in Central Arizona but sorely needed due to the rapidity of habitat loss and degradation in the region. Since this time, and since the iterative convenings of the ROSS, many collaborative networks have formed in Central Arizona, illustrating a change in the environment for collaboration here. As of 2022, some of the primary collaboratives in Central Arizona, aside from CAZCA itself, include the Lower Gila Collaborative, Rio Reimagined, and the North Valley Outdoor Network. Organizations such as White Tank Mountains Conservancy and McDowell Sonoran Conservancy have also become leaders in collaborative work, and both are CAZCA partners.

3.3. *Great Milkweed Grow Out*

Great Milkweed Grow Out (GMGO) was well positioned at its inception to collaborate with federal agencies and monarch conservation organizations to carry out projects that would directly benefit monarch conservation. Monarch butterflies are captivating and have inspired conservation action within the community and interest from many potential partners. From 2016–2019, GMGO's propagation work focused on milkweeds (*Asclepias* spp.). GMGO collaborated with the federal Bureau of Land Management (BLM) to grow and distribute milkweeds and collect seeds from wild populations. In that time, more than 6000 milkweeds were propagated and distributed to agencies, non-profits, and members of the general public. Milkweeds quickly sold out at plant sales and were popular giveaways at events. GMGO developed growing protocols for three species of native milkweeds from wild-collected seed. GMGO successfully demonstrated the demand for native milkweeds and shared these growing protocols. Since 2016, local wholesale growers have begun to grow and offer more native milkweed taxa, making milkweeds and monarch habitat even more prevalent within the region.

In the following years, GMGO broadened the scope of propagation, outreach, and research to include other pollinators that are also facing significant population declines [53,54] and thus to grow and distribute plants beyond milkweeds. GMGO has propagated and developed replicable propagation protocols for a range of native plants valuable to pollinators, including *Aristolochia watsonii*, *Acourtia wrightii*, *Cephalanthus occidentalis*, *Conoclinium greggii*, and *Ageratum corymbosum*. Since 2019, GMGO has propagated and distributed more than 1500 non-milkweed plants and continues to grow at least 1500 milkweeds per year. To further increase plant material availability, GMGO has worked with BLM to collect milkweed and other native nectar plant seeds under BLM's Seeds of Success Protocol. GMGO's seed collections and propagation efforts were recognized on a national scale in the 2018 National Seed Strategy Making Progress Report [46].

In addition to distributing plants, GMGO creates pollinator gardens in schools and community centers in the Phoenix area, planting 22 pollinator gardens since 2016. These pollinator gardens are a collaboration with BLM and were featured in their 2015–2020 National Seed Strategy Progress Report [55]. GMGO has also donated plants to more than 25 partner organizations to facilitate habitat creation and has given more than 20 presentations to community groups, schools, non-profit organizations, and corporate groups. Three larger areas of habitat have been created in partnership with the US Fish and Wildlife Service's (USFWS) Partners for Fish and Wildlife Program.

GMGO has significantly contributed to our understanding of milkweeds and monarch behavior and ecology through a variety of research projects. In 2020, GMGO published a study on the egg-laying preference and larval performance of monarch butterflies on native milkweed taxa [44]. The research showed higher preference and survival for one native species (*Asclepias angustifolia*) compared to another (*A. linaria*). This directly influenced recommendations for plantings within the region and influenced the focus of GMGO milkweed propagation. GMGO has also researched the pollinators, parasites, and predators that visit milkweed and could benefit milkweed populations or endanger growing monarch caterpillars (manuscript in progress). In 2018, GMGO initiated a number of research projects still ongoing in collaboration with the University of Arizona on milkweed physiology relating to host–pollinator interactions and climate shifts in order to guide habitat plantings in the future. In February 2022, GMGO launched a community science project in collaboration with the USA National Phenology Network to gather data on monarch behavior and milkweed phenology in the winter months in Arizona [56]. This will also contribute to community engagement by offering monarch enthusiasts a chance to be more involved.

Through all these activities, GMGO has worked closely with other organizations interested in monarch and pollinator conservation to have the greatest impact. Since 2015, DBG has been working with Monarch Joint Venture (MJV), an international partnership of agencies, NGOs, and academic institutions. As a partner in MJV, DBG has benefited from significant knowledge transfer and a framework for conservation priorities [57]. Statewide, GMGO has been active in the Arizona Monarch Collaborative (AMC). Created in 2019, AMC has brought together approximately 70 different partners across the state [58]. Partners include state agencies, local representatives of federal agencies, non-profits, energy companies, tribes, and universities. DBG has been active since the start, providing expertise on monarchs in Arizona and the plants that support them. As of March 2022, DBG has one staff member on the steering committee and staff members leading the plant materials and restoration and the research and monitoring committees. DBG staff serve as experts within this group and have shared growing protocols and knowledge of local plant species and monarch biology with AMC partners. This has led to the development of conservation goals for the state in support of the Western Association of Fish and Wildlife Agencies' Monarch Conservation Plan [59] and a variety of resources being made available on the website to wide audiences [60].

3.4. Rare, Threatened, and Endangered Species Conservation

DBG's participation and leadership role within the Center for Plant Conservation (CPC) has led to the conservation of numerous imperiled species. The DBG seed bank currently houses over 5300 seed accessions, 2500 from plants of conservation concern. Since 1984, DBG has made 708 seed collections of 52 CPC National Collection species and currently maintains 566 accessions in the DBG seed bank. During that period, over 400 duplicates of these collections have been backed up at the United States Department of Agriculture's National Laboratory for Germplasm Resource Preservation in Ft. Collins, Colorado, through an agreement with CPC. Since partnering with North American Orchid Conservation Center (NAOCC), DBG has contributed 38 accessions of 13 taxa towards the goal of banking seeds of all orchid species of the Southwest. In addition, 30 root samples have been provided to the Smithsonian Environmental Research Center (SERC), which has led to the successful identification of several fungal symbionts associated with these orchids.

In support of both CPC and NAOCC, DBG is working under a USFWS-funded grant to develop methods to propagate the endangered orchid *Spiranthes delitescens* (the Canelo Hills ladies' tresses orchid) for use in future seed amplification and restoration projects. In 2016 and 2021, seeds were collected from the two extant populations and have been banked and used to successfully propagate plants in vitro. These seedlings have since been moved into potting media for further growth to maturity. In addition, SERC has been able to culture and identify its symbiotic fungus. This fungus can then be used to inoculate seeds to allow for plants to be propagated using common horticultural methods [61].

Due to the cryptic nature of many native terrestrial orchids, plants can be difficult to locate in the wild given their size, shape, and ability to blend with surrounding vegetation. Added to this, many of the rarer species, such as *S. delitescens*, exist in small communities. To aid in locating these plants, DBG, with the support of USFWS, has enlisted a professional dog trainer experienced with using dogs for ecological purposes. The goal is to train dogs to detect these orchids in situ to locate additional plants at known sites and relocate populations at historical sites. In June 2022, collections of 13 living plants were made of a congener of *S. delitescens*, *S. infernalis* (Ash Meadows ladies' tresses), currently being used as a proxy for training the dogs off-site. Early results have shown that the dogs are able to detect this related species which should lead to successfully locating the target orchids in the wild. These projects would not be possible without cooperation between federal agencies, private landowners, and private businesses.

In 2018, DBG was contracted by the Arizona Department of Transportation (ADOT) to conduct a salvage and reintroduction of the endangered *Echinocereus arizonicus* (Arizona hedgehog cactus) to make way for a new bridge construction over Pinto Creek, near Superior, Arizona. The salvage consisted of 34 mother plants that were collected from the construction site surrounding Pinto Creek. Due to the steep, rocky terrain, the salvage required extreme physical effort, including rock climbing and rappelling, and care in delicately extracting the plants from their habitat. Each cactus was documented and transported back to the Garden, where they were cared for and allowed to rehabilitate in the Hazel Hare Center for Plant Science facilities. While housed at DBG, the plants were hand-pollinated, and seeds were collected for banking, testing, and propagation. This resulted in 12 additional seed accessions of known parentage containing thousands of seeds, hundreds of seedlings produced, and a collection of cuttings still maintained at DBG.

In March 2022, after the completion of the bridge, the same team of staff returned to the heavily impacted construction site to reintroduce 61 plants consisting of mostly original mother plants along with cuttings propagated from those same plants. Care was taken to transplant the cacti within the right microsites. They were given ID tags, measured, and provided with supplemental water. Under the agreement, DBG will monitor these plants for two years and augment the population as necessary using plant material still maintained at DBG. Although funding was provided by ADOT, this project was a partnership between DBG, federal agencies, and private businesses and has been an incredible achievement in collaborative conservation.

DBG has also had considerable success working with private landowners. This collaboration is particularly important since federally listed plants are not granted the same protection as animals on private lands under the Endangered Species Act [62]. As part of the CPC, NAOCC, and USFWS collaboration to conserve *S. delitescens,* DBG has been working with private landowners to locate new and historic populations. The orchid was historically known to only exist in four locations in Arizona. Three locations were on privately owned cattle ranches and another on a preserve managed by the Nature Conservancy. DBG was able to assess these historical populations by working with private landowners.

The participation of the Garden in the IUCN Cactus and Succulent Specialist Group as part of the Species Survival Commission has allowed for greater scientific and media attention to the dire status of cacti and succulents and the threats to their populations. DBG scientists worked with other cactus succulent specialists around the world to complete Red List Assessments. Of the 1478 evaluated species of cacti, nearly 31% were found to be at risk of extinction [63]. DBG is currently working with other members of the Cactus and Succulent Specialist group to assess 300 species of aloes and yuccas, with most of the assessments completed and published by the Red List [64].

4. Discussion

4.1. EcoFlora

The EcoFlora program is a compelling and dynamic tool that illustrates the importance of science communication and community science for collaborative conservation at any scale. The Metro Phoenix EcoFlora (MPE) is an inherently collaborative project, from its inception at New York Botanical Garden, to its shared implementation by partner gardens, to working with partner organizations and community scientists. The objectives of both the joint EcoFlora program and the local MPE project are largely achieved through collaboration and engagement, along with the provision of accessible resources and open-source data. Botanical gardens can play a unique role and engage and support projects that are invested in collaborative conservation efforts in numerous ways.

The established communication channels and built-in audiences with DBG, Central Arizona Conservation Alliance (CAZCA), and iNaturalist provided the opportunity for initial awareness raising as well as supporting communication, sharing resources, and sustaining engagement in the MPE project (Figure 6). DBG, specifically, is a well-known organization that has strong support and enthusiasm from the public and is visited by nearly half-a-million people on an annual basis. MPE's association with DBG is in itself a draw for the public to join the project as well as attend classes and events. In-house marketing and design professionals at botanical gardens can assist with broad engagement and graphic communications. The Marketing Communications team at DBG provides essential support for and promotion of MPE, including project branding for all five partner gardens and the national program, designing collateral and materials, website design and hosting, and email and social communications.

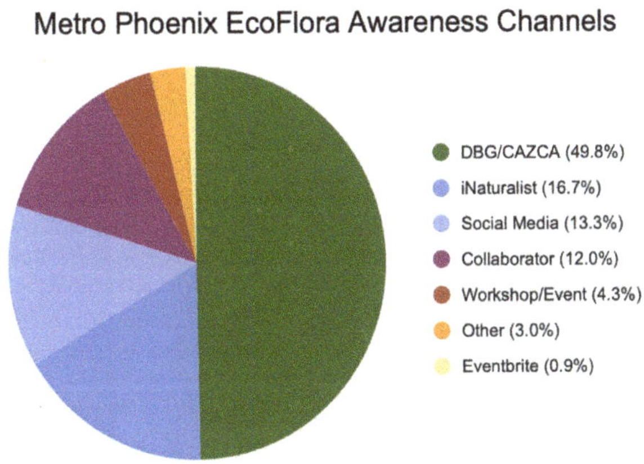

Figure 6. Channels of MPE project awareness as of 10 March 2022 according to MPE newsletter responses.

Events, classes, and resources created by MPE are free and open to project members and the public, with an overall result of increased awareness and knowledge of biodiversity and plant life. DBG provides the space for events and classes, requiring fewer resources, lessening the burden on MPE to find, schedule, reserve, and finance spaces outside of DBG. MPE creates and collaborates on resources that can be used by other projects and organizations. For example, MPE worked on educational videos with McDowell Sonoran Conservancy's Conservancy Kids program, teaching younger audiences about biodiversity and observing nature. Numerous educational resources that have been created by MPE are available through the iNaturalist project journal. DBG email lists and social media provide a way to share these resources widely.

Botanical gardens are uniquely situated to combine modern observations with existing floristic data. A deliverable of the EcoFlora project is to synthesize iNaturalist observations with existing herbarium records through Symbiota (https://symbiota.org/ accessed on 10 March 2022), the biodiversity data management software that hosts the SEINet portal. A collective Symbiota portal will be created for the EcoFlora partner gardens, linking out to checklists and information for each garden and the flora that can be found in their respective locations. Utilizing SEINet and the herbarium resources and expertise at DBG, MPE has identified 40 non-cultivated plant species that have not been previously documented in the metro Phoenix area. Furthermore, iNaturalist can be used to compliment herbarium vouchers by adding a QR code to labels. The QR code can link to iNaturalist observations that can include photos of the live specimen and other information. iNaturalist, Symbiota, and SEINet are open-source platforms, and botanical gardens and scientists around the world can access and use EcoFlora project data and information, contributing to conservation efforts locally and at large.

Finally, botanical gardens and other organizations across the United States have expressed interest in applying the EcoFlora model, while others have already begun to do so, adding to local and national conservation efforts. MPE provided the foundational materials and support for the Maricopa County Parks and Recreation's Eco-BLITZ program, which focuses on observations in Maricopa parks (Arizona, USA). The project also inspired the Fire Followers program from the California Native Plant Society (CNPS, California, USA). Jose Esparza, Community Science Coordinator for CNPS, states: "The Metro Phoenix EcoFlora project on iNaturalist was truly inspirational to me during the launch of our own CA Fire Followers Project. It is amazing to see other participatory science projects on iNaturalist be community-driven and characterized by placed-based knowledge and social learning, collective action and empowerment." In June 2022, the EcoFloras of North America created an instructional toolkit [65] that can be shared with botanical gardens interested in implementing an EcoFlora or similar program. The toolkit is specifically designed with the capabilities and resources that botanical gardens can provide for an EcoFlora project.

4.2. Central Arizona Conservation Alliance

The Central Arizona Conservation Alliance (CAZCA) provides insight into the strategic benefits of botanical garden leadership in collective impact backbone organizations and has illuminated the challenges of this kind of collaborative conservation as well.

First, while some botanical gardens, such as DBG, have strong conservation commitments [28], they can nonetheless serve as neutral facilitators and provide neutral convening spaces for partners that might otherwise hesitate to collaborate with one another. This is ideal for the collaborative process, and such resources can assist organizations in building social capital and supporting essential governance formation for the collective effort [8,65]. They also provide welcoming and non-threatening spaces for new collaborators, particularly those representing groups that might not traditionally be involved in conservation work but whose participation is now understood to be essential [66]. Such neutral spaces strengthen collaborations and their related conservation outcomes through the support of inclusive decision making and resulting shared buy-in [67]. Gardens also have physical infrastructures for convenings and uniquely attractive spaces for large and small gatherings which implicitly encourage stakeholder involvement [68]. The organization of large, diverse groups in the creation of CAZCA's Regional Open Space Strategy, Greenprint, and Conservation Opportunity Areas demonstrate this strength.

Botanical gardens can also serve as a natural bridge between conservation professionals and the public because the gardens themselves are designed as spaces for public immersion and connection with nature, and many gardens have internal marketing, communication, and outreach expertise [69]. In terms of collaborative efforts, CAZCA has illustrated the benefit of a botanical garden serving as a backbone organization through its successful awareness-raising campaigns and public engagement efforts.

Finally, botanical gardens have resources from their admissions revenues, grants, and donor support to provide the needed capacity for collaborative conservation efforts. Dedicated staff and expertise have been found to have strong positive impacts on collaborative, landscape-scale outcomes [10]. When considering the conservation challenges created by plant invisibility, having garden staff in leadership roles in collaborative efforts can help elevate and communicate the importance of plants in biodiversity work-at-scale, increasing the longevity of positive conservation outcomes [70]. CAZCA's regional impacts on the collaborative environment as well as its impacts on plant-focused conservation efforts have underlined these advantages.

4.3. Great Milkweed Grow Out

The success of Great Milkweed Grow Out (GMGO) as a conservation initiative is due to a variety of factors, such as dedicated staff and volunteers, incorporation of a charismatic animal, and a wide variety of collaborators in our state and region. Some of these are due to GMGO's positioning at a botanical garden.

While monarch and pollinator conservation programs exist in many communities and institutions on many scales, one factor that has made GMGO successful is the plant expertise provided by DBG. With native plants at the heart of monarch conservation, this expertise is essential for actions such as seed germination, planting schedules, plant mixes, responsible native seed collection, milkweed diversity, and large-scale propagation [71–73]. With a staff that includes horticulturalists, propagators, research botanists, and conservation experts, DBG is uniquely positioned among its partner organizations to provide this information. Expertise in pollinator and butterfly biology was also available due to DBG's investment in the Butterfly Exhibit and having a full-time butterfly expert on staff to oversee the health and welfare of the butterflies. Botanical gardens often hold a wealth of knowledge on many different ecological topics, and collaborative conservation efforts can benefit from this.

The blend of floral and faunal focus has also allowed the program to benefit from varied funding sources. Funding from botanical and horticultural interests have funded propagation, internships, and seed collection. Organizations and programs with an interest in monarchs and pollinators have funded school gardens, habitat plantings, and research. Portions of the program are funded by plant and seed sales through DBG's bi-annual plant sale and Garden Shop, which attract gardeners from the greater Phoenix area.

GMGO has been able to effectively share important conservation messages that inspire action by reaching a wide variety of audiences through DBG. GMGO gives DBG the opportunity to share impactful messages about the importance of native plants with guests that connect more with monarchs and other animals than with plants. This is an audience DBG may not have reached otherwise. Similarly, DBG attracts many avid gardeners that seek out unique cacti, succulents, and other desert plants and admire their beauty. By sharing the importance of monarch conservation, many dedicated gardeners have begun seeking milkweed and other nectar plants to support monarchs and pollinators. Pollinators are bridging the gap between flora and fauna and inspiring collaboration and action.

The second major factor in GMGO's success in collaborative conservation has been the resources available at DBG. The most important of these resources is their volunteer force. DBG hosts and manages a large, dedicated volunteer group, generally maintaining close to 800 active volunteers that assist with a wide variety of programs. GMGO worked to engage these volunteers from the beginning of the initiative, and most propagation tasks are carried out by dedicated DBG volunteers led by staff. Since 2016, GMGO has worked with over 330 volunteers contributing more than 4500 h to propagation, outreach, and research projects (as of February 2022). GMGO would not be as successful without this volunteer base.

Another set of key factors is space and facilities. Locations not actively cultivated or open to the public exist within DBG's footprint that can be used for propagation and experimental plots, including greenhouse and garden space, and butterfly husbandry facilities. GMGO uses roughly 6000 square meters of outdoor space for propagation of plants in one-gallon plots. GMGO also built experimental beds to test hypotheses about monarch and queen visitation to different milkweed species [44] and characterize the wide variety of beneficial insects supported by milkweeds. DBG's property size also allowed for the development of a $\frac{1}{4}$-acre fenced experimental plot to test the effects of drought due to climate change on milkweeds. This area and another nearby also serve as seed amplification plots for harvesting known-source milkweed seeds to share with various partners for habitat restoration. Finally, the Butterfly Exhibit has served as a space for controlled larval rearing experiments and studies of behavior.

The last major factor that allows for success is the name recognition of DBG. DBG is recognized throughout the region as an expert in local natural resources. This lends credibility to GMGO's conservation actions, and facilitates outreach, as most has been the result of word-of-mouth contacts. GMGO did not have to spend time and effort seeking opportunities to interact with the community or schools that wanted a butterfly garden.

4.4. *Rare, Threatened, and Endangered Species Conservation*

Plant conservationists recognize that saving plants cannot be achieved in isolation. The threats to habitats and ecosystems are simply too great and one institution alone cannot single-handedly keep up with the increasing rate of species loss. Conservation efforts at DBG have benefited greatly by working collaboratively alongside government agencies, businesses, and private landowners.

The success of collaborative conservation is partially due to the extensive resources at DBG and the status of DBG as a non-governmental organization (NGO). DBG has made considerable investment in plant care facilities, such as the Hazel Hare Center for Plant Science, and in the development of a robust research and conservation department with extensive knowledge and expertise, which has benefited the conservation of rare species. The position of Conservation Collections Manager at DBG, for example, oversees the seed bank, rare plants in cultivation, and conducts research on rare and endangered plants. This position is also tasked with obtaining and managing permits, acting as a conduit between the Garden and outside entities.

For decades, DBG has partnered with the US Fish and Wildlife Service (USFWS) and the Bureau of Land Management in the conservation of endangered plants and protection of rare species. Since agencies often have limited staff, resources, and expertise for managing or mitigating the loss of rare species, they have turned to outside organizations, such as botanical gardens, to fulfill these requirements. As demonstrated in the case studies, DBG has had a history of successful partnerships with federal and state agencies, and without the assistance of botanical institutions, the capacity for conservation by federal agencies would be greatly reduced.

When rare plants occur on private lands, collaboration with landowners is necessary. DBG's status as a non-governmental organization allows for a bridge between federal or state governments and private landowners. Landowners may be skeptical of the motives of government agencies, resulting in distrust and unwillingness to cooperate in conservation efforts. The reputation of DBG as a respected institution has been valuable in opening lines of communication with landowners. In several cases, DBG has been able to access private land to monitor imperiled species, where the same luxury may not have been afforded to government officials.

4.5. Challenges to Collaboration

While there are many demonstrated benefits of botanical garden leadership and involvement in collaborative conservation efforts, there are also a variety of challenges that should be taken into consideration. First, collaborative conservation efforts require resources in the form of funding and staff capacity, particularly for any institutions attempting to lead or serve as a backbone organization [10]. Leading and coordinating successful collaborative conservation efforts is complicated, dynamic, and difficult; sufficient resources over the long-term are essential [1,74]. In the case of gardens, staff capacity may also be shared between collaborative responsibilities and duties related to garden operations, even for those staff members with positions primarily focused on collaborative efforts. Collaborative conservation may have the benefit of combining resources across institutions and can be a great way to leverage capacity and funds across institutions. Gardens themselves may have more flexibility than agencies or other institutions to accept types of funding or add staff positions to their rosters. Gardens may also have resources through alternate streams of funding that most other partner organizations cannot take advantage of via ticket and membership sales, event-space rentals, etc. However, this kind of revenue, widely related to tourism, whether local, domestic, or international, can be heavily impacted in times of economic strain or uncertainty [75]. Without sufficient, long-term funding, collaborations are unlikely to be successful.

Additionally, collaboration with multiple agencies and private landowners takes time and sometimes negotiation skills to navigate conflict. Conservation efforts may be time-sensitive due to seasonality of plant phenophases or schedules of planned development. Working on federal, state, and municipal lands but not being a related agency requires the acquisition of permits, which can take months or longer to process. In addition, culturally diverse, minoritized, and under-resourced communities are understandably more likely to engage with community-based organizations than with scientific and informal science-education institutions and organizations, and understanding the values, priorities, and boundaries of communities should be prioritized to ensure their autonomy and balance of power [76].

Collaborative efforts are, by nature, more complex than conservation initiatives that are primarily or entirely designed and managed by a single entity. They require the cooperation of at least two partners and the coordination of their personnel and resources. This complexity leads to increased need for capacity and related funding requirements, the potential need for long-term work beyond the time horizons of typical funders, and additional skills needed for collaborative teams related to trust-building and working across institutional boundaries. These are challenges that must be taken into account but which can be overcome.

5. Conclusions

There are many benefits to collaborative conservation with botanical garden partners. Botanical gardens have access to an immense amount of botanical knowledge that is not widely available elsewhere. Many botanical gardens are non-profit institutions that have unique, diverse revenue streams via admissions, committed donors and members, and a variety of public and private grants [77–79]. Botanical gardens have the ability to secure funding for a wide variety of project types, including research and restoration activities, educational programming, and trail renovation. Diversity of funding and project types makes these institutions more resilient to funding challenges and also creates many opportunities for partnership. Many botanical gardens also utilize volunteers to assist with conservation projects [80]. Botanical gardens often house their own marketing, communication, and design professionals who can provide effective tools for outreach and conservation efforts. [74].

For botanical gardens that are looking to lead conservation, these case studies provide key takeaways. First, forming relationships with local and federal agencies, NGOs, and academic institutions is imperative to the success of conservation programs. This is also something important for agencies to keep in mind—to look to the expertise and experience that botanical gardens may hold. Second, ideas for impactful conservation programs can emerge from the creativity of staff below leadership level. Staff imagination and ambition, with institutional support, can lead to impactful and profitable programs. Investing in staff who have relevant experience, expertise, and relationships is important to leading collaborative conservation. Third, careful planning around collaborative projects is necessary, even when the need for collaborative efforts is clear. It is imperative that realistic timelines are developed, long-term funding needs are addressed, and the well-being and collaborative skills of backbone staff and active partners are inventoried and maintained. Conservation programs involve direct connections to community and partner groups that bring value and recognition to their host institutions and, with proper internal support and respect, provide ample opportunities for the pursuit of further institutional as well as collaborative goals.

The role of botanical gardens in conservation has evolved over time, and it is clear that these institutions are essential components of the effort to slow and stop the rapid loss of global biodiversity through the direct protection of plant species and the species that rely on them [14]. As the importance of collaboration in conservation is becoming more clear, botanical gardens have and will continue to serve as leaders in collective efforts to manage natural resources. Understanding the benefits and challenges related to this leadership, as reflected in the case studies explored here, can assist botanical gardens and partners of botanical gardens in starting and maintaining the successful collaborations that are increasingly needed to preserve and cultivate healthy ecosystems.

Author Contributions: Conceptualization, A.B.R., K.V.P., N.A.M. and J.D.; methodology, A.B.R., K.V.P., N.A.M., J.D. and S.A.B.; data curation, A.B.R., K.V.P., N.A.M., J.D. and S.A.B.; writing—original draft preparation, A.B.R., K.V.P., N.A.M., J.D. and S.A.B.; writing—review and editing, A.B.R., K.V.P., N.A.M., J.D. and S.A.B.; project administration, A.B.R., K.V.P., N.A.M., J.D. and S.A.B.; funding acquisition, A.B.R., K.V.P., N.A.M. and S.A.B. All authors have read and agreed to the published version of the manuscript.

Funding: CAZCA was funded by the Nina Mason Pulliam Charitable Trust. EcoFlora was funded by the Institute of Museum and Library Services, grant number MG-70-19-0057-19. GMGO was funded by Desert Botanical Garden, Bureau of Land Management, US Fish and Wildlife Service, Gila River Indian Community, Stanley Smith Horticultural Trust, Arthur L. and Elaine V. Johnson Foundation, and a private donor. DBG's plant conservation is funded by the Bureau of Land Management, US Fish and Wildlife, and private donors.

Institutional Review Board Statement: Not applicable.

Data Availability Statement: Not applicable.

Acknowledgments: For EcoFlora, we thank New York Botanical Garden as the founding EcoFlora organization; Desert Botanical Garden as the host institution; the Central Arizona Conservation Alliance as the host organization; Chicago Botanic Garden, Denver Botanic Gardens, and Marie Selby Botanical Gardens as supporting partner organizations; the numerous contributing project members and collaborators; Volunteers in the Garden (VIGs); and the Marketing Communications team at Desert Botanical Garden. For CAZCA, we thank all of our partner organizations, volunteers, and funders for their contributions of time, funding, and expertise. For Great Milkweed Grow Out and rare plant conservation, we thank all of our volunteers, partners, and funders for their support and assistance.

Conflicts of Interest: The authors declare no conflict of interest. The funders had no role in the design of the study; in the collection, analyses, or interpretation of data; in the writing of the manuscript, or in the decision to publish the results.

Appendix A. CAZCA Partners and Collaborators in 2018

Desert Botanical Garden
Arizona Alliance for Liveable Communities
Arizona Association for Environmental Education
Arizona Columbine Garden Club
Arizona Center for Nature Conservation
Arizona Department of Transportation
Arizona Department of Water Resources
Arizona Game and Fish Department
Arizona Native Plant Society
Arizona Parks and Recreation Association
Sonoran Desert Museum
ASU—Arizona Sustainable Cities Network
ASU—Biomimicry Center
ASU—CAP LTER
ASU—Design School
ASU—Global Institute of Sustainability
ASU—School of Community Resources and Development
ASU—School of Life Sciences
ASU—UREx Sustainable Research Network
Audubon Arizona
B3.8
Bureau of Land Management
Bureau of Reclamation
Catalyst Collective
Center for Biodiversity Outcomes
Center for the Future of Arizona
City of Buckeye Community Services Department
City of Glendale Community Services Department
City of Peoria Community Services Department
City of Phoenix Parks and Recreation Department
City of Scottsdale Parks and Recreation Department
Cultivate South Phoenix—Spaces of Opportunity
Desert Foothills Land Trust
Desert Foothills Community Foundation, Desert Awareness Committee
Friends of Daisy Mountain Trails
International Union for Conservation of Nature
Keep Phoenix Beautiful
Local First Arizona

Logan Simpson & Associates
Maricopa Association of Governments
Maricopa County Parks and Recreation Department
Maricopa Trail and Parks Foundation
McDowell Sonoran Conservancy
National Park Service—Saguaro National Park
North Mountain Visitor Center
Northern Arizona University
Okanogan Trail Construction, Inc.
Phoenix College
Phoenix Fire Department
Phoenix Parks Foundation
Phoenix Union High School District
Phoenix Weedwackers
Pima County Natural Resources, Parks, and Recreation
Pinal County Open Space and Trails
PLANet
Plant Atlas Project of Arizona
Salt River Pima—Maricopa Indian Community
Save Our Mountains Foundation
Signature Botanica
Sonoran Institute
South Mountain Environmental Education Center
Southwest Seed Partnership
Southwest Society of Botanical Artists
Superstition Area Land Trust
Tovrea Carraro Society
The Center for Native and Urban Wildlife
The Nature Conservancy
The Phoenix Mountains Preservation Council
The Trust for Public Land
USGS—Western Ecological Research Center
USFS—Tonto National Forest
Vitalyst Foundation
White Tank Mountains Conservancy

Appendix B. List of Participating Organizations in the Stakeholder Consultation Meetings for the Conservation Opportunity Area Identification

Arizona Army National Guard
Arizona Game and Fish Department
Arizona Office of Tourism
APS
Arizona State Parks and Trails OHV
Arizona Water Company
Arizona Wilderness Coalition
ASU School of Community Resources and Development
ASU School of Landscape Design
ASU Sustainable Cities
Bureau of Land Management
Circle G Development
City of Apache Junction
City of Avondale
City of Buckeye
City of Mesa

City of Peoria
City of Phoenix
Copper State Consulting Group
Desert Foothills Mountain Bike Association
El Dorado Holdings
Goodyear Recreation Board
GPEC
Maricopa County Planning
Maricopa Farm Bureau
Maricopa Trails and Park Foundation
MBAA
Town of Cave Creek
Town of Surprise
State Historic Preservation Office
Superstition Area Land Trust
Valley Partnership
Vitalyst Health Foundation
Retired Supervisor of Maricopa County (Individual)
Landscape Architect (Individual)

References

1. Wilkins, K.; Pejchar, L.; Carroll, S.L.; Jones, M.S.; Walker, S.E.; Shinbrot, X.A.; Huayhuaca, C.; Fernández-Giménez, M.E.; Reid, R.S. Collaborative Conservation in the United States: A Review of Motivations, Goals, and Outcomes. *Biol. Conserv.* **2021**, *259*, 109165. [CrossRef]
2. Berkes, F.; Colding, J.; Folke, C. Rediscovery of Traditional Ecological Knowledge as Adaptive Management. *Ecol. Appl.* **2000**, *10*, 1251–1262. [CrossRef]
3. Brick, P.; Snow, D.; Bates, S.F.; Kemmis, D. (Eds.) *Across the Great Divide: Explorations in Collaborative Conservation and the American West*; Island Press: Washington, DC, USA, 2001.
4. Knight, R.L.; White, C. (Eds.) *Conservation for a New Generation: Redefining Natural Resources Management*; Island Press: Washington, DC, USA, 2009.
5. Nelson, F.; Parrish, J. Developing New Models for Collaboration in Conservation. Available online: https://ssir.org/articles/entry/developing_new_models_for_collaboration_in_conservation (accessed on 14 March 2022).
6. Franks, J.R.; Emery, S.B. Incentivising Collaborative Conservation: Lessons from Existing Environmental Stewardship Scheme Options. *Land Use Policy* **2013**, *30*, 847–862. [CrossRef]
7. Wyborn, C.; Bixler, R.P. Collaboration and Nested Environmental Governance: Scale Dependency, Scale Framing, and Cross-Scale Interactions in Collaborative Conservation. *J. Environ. Manag.* **2013**, *123*, 58–67. [CrossRef]
8. Cox, M.; Arnold, G.; Tomás, S.V. A Review of Design Principles for Community-Based Natural Resource Management. *Ecol. Soc.* **2010**, *15*, 38. [CrossRef]
9. Ostrom, E. *Governing the Commons: The Evolution of Institutions for Collective Action*; Cambridge University Press: Cambridge, UK, 1990.
10. Kania, J.; Kramer, M. Collective Impact. Available online: https://ssir.org/articles/entry/collective_impact (accessed on 10 February 2022).
11. Jackson, P.W.; Bridge, B.; Dennis, F.; Leadlay, E.; Hobson, C.; Holland, F.; Pendry, T.; Skilton, J.; Sutherland, L.; Willison, J.; et al. An International Review of the Ex Situ Plant Collections of the Botanic Gardens of the World. *Bot. Gard. Conserv. News* **2001**, *3*, 22–33.
12. Maunder, M.; Higgens, S.; Culham, A. The Effectiveness of Botanic Garden Collections in Supporting Plant Conservation: A European Case Study. *Biodivers. Conserv.* **2001**, *10*, 383–401. [CrossRef]
13. Mounce, R.; Smith, P.; Brockington, S. Ex Situ Conservation of Plant Diversity in the World's Botanic Gardens. *Nat. Plants* **2017**, *3*, 795–802. [CrossRef]
14. Chen, G.; Sun, W. The Role of Botanical Gardens in Scientific Research, Conservation, and Citizen Science. *Plant Divers.* **2018**, *40*, 181–188. [CrossRef]
15. O'Donnell, K.; Sharrock, S. The Contribution of Botanic Gardens to Ex Situ Conservation through Seed Banking. *Plant Divers.* **2017**, *39*, 373–378. [CrossRef]
16. Breman, E.; Ballesteros, D.; Castillo-Lorenzo, E.; Cockel, C.; Dickie, J.; Faruk, A.; O'Donnell, K.; Offord, C.A.; Pironon, S.; Sharrock, S.; et al. Plant Diversity Conservation Challenges and Prospects—The Perspective of Botanic Gardens and the Millennium Seed Bank. *Plants* **2021**, *10*, 2371. [CrossRef] [PubMed]

17. Donaldson, J.S. Botanic Gardens Science for Conservation and Global Change. *Trends Plant Sci.* **2009**, *14*, 608–613. [CrossRef] [PubMed]
18. Westwood, M.; Cavender, N.; Meyer, A.; Smith, P. Botanic Garden Solutions to the Plant Extinction Crisis. *Plants People Planet* **2021**, *3*, 22–32. [CrossRef]
19. Chen, J.; Cannon, C.H.; Hu, H. Tropical Botanical Gardens: At the in Situ Ecosystem Management Frontier. *Trends Plant Sci.* **2009**, *14*, 584–589. [CrossRef] [PubMed]
20. Cannon, C.H.; Kua, C.-S. Botanic Gardens Should Lead the Way to Create a "Garden Earth" in the Anthropocene. *Plant Divers.* **2017**, *39*, 331–337. [CrossRef] [PubMed]
21. He, H.; Chen, J. Educational and Enjoyment Benefits of Visitor Education Centers at Botanical Gardens. *Biol. Conserv.* **2012**, *149*, 103–112. [CrossRef]
22. Sellmann, D.; Bogner, F.X. Climate Change Education: Quantitatively Assessing the Impact of a Botanical Garden as an Informal Learning Environment. *Environ. Educ. Res.* **2013**, *19*, 415–429. [CrossRef]
23. Kahtz, A.W. Impact of Environmental Education Classes at Missouri Botanical Garden on Attitude and Knowledge Change of Elementary School Children. *HortTechnology* **1995**, *5*, 338–340. [CrossRef]
24. Zelenika, I.; Moreau, T.; Lane, O.; Zhao, J. Sustainability Education in a Botanical Garden Promotes Environmental Knowledge, Attitudes and Willingness to Act. *Environ. Educ. Res.* **2018**, *24*, 1581–1596. [CrossRef]
25. Havens, K.; Vitt, P.; Masi, S. Citizen Science on a Local Scale: The Plants of Concern Program. *Front. Ecol. Environ.* **2012**, *10*, 321–323. [CrossRef]
26. Raschke, A.B.; Davis, J.; Quiroz, A. The Central Arizona Conservation Alliance Programs: Use of Social Media and App-Supported Community Science for Landscape-Scale Habitat Restoration, Governance Support, and Community Resilience-Building. *Land* **2022**, *11*, 137. [CrossRef]
27. Desert Botanical Garden Living Collections. Available online: https://livingcollections.org/dbg/Home.aspx (accessed on 12 March 2022).
28. Desert Botanical Garden. *Strategic Plan: 2019–2023*; Desert Botanical Garden: Phoenix, AZ, USA, 2019.
29. Alcock, I.; White, M.P.; Pahl, S.; Duarte-Davidson, R.; Fleming, L.E. Associations between Pro-Environmental Behaviour and Neighbourhood Nature, Nature Visit Frequency and Nature Appreciation: Evidence from a Nationally Representative Survey in England. *Environ. Int.* **2020**, *136*, 105441. [CrossRef] [PubMed]
30. Liu, Z.; He, C.; Wu, J. The Relationship between Habitat Loss and Fragmentation during Urbanization: An Empirical Evaluation from 16 World Cities. *PLoS ONE* **2016**, *11*, e0154613. [CrossRef] [PubMed]
31. MAG Fast Facts—Administration. Available online: https://azmag.gov/About-Us/Divisions/Administration-Division/MAG-Fast-Facts-Administration (accessed on 29 December 2021).
32. Parkes, M.W.; Morrison, K.E.; Bunch, M.J.; Hallström, L.K.; Neudoerffer, R.C.; Venema, H.D.; Waltner-Toews, D. Towards Integrated Governance for Water, Health and Social–Ecological Systems: The Watershed Governance Prism. *Glob. Environ. Change* **2010**, *20*, 693–704. [CrossRef]
33. Beute, S.; Arndt, L. The Regional Open Space Strategy for Maricopa County 2018. Beute, S., Arndt, L., Eds.; In *The Regional Open Space Strategy for Maricopa County*; Central Arizona Conservation Alliance: Phoenix, AZ, USA, 2018.
34. Schultz, C.B.; Brown, L.M.; Pelton, E.; Crone, E.E. Citizen Science Monitoring Demonstrates Dramatic Declines of Monarch Butterflies in Western North America. *Biol. Conserv.* **2017**, *214*, 343–346. [CrossRef]
35. Thogmartin, W.E.; Wiederholt, R.; Oberhauser, K.; Drum, R.G.; Diffendorfer, J.E.; Altizer, S.; Taylor, O.R.; Pleasants, J.; Semmens, D.; Semmens, B.; et al. Monarch Butterfly Population Decline in North America: Identifying the Threatening Processes. *R. Soc. Open Sci.* **2017**, *4*, 170760. [CrossRef] [PubMed]
36. Pelton, E.M.; Schultz, C.B.; Jepsen, S.J.; Black, S.H.; Crone, E.E. Western Monarch Population Plummets: Status, Probable Causes, and Recommended Conservation Actions. *Front. Ecol. Evol.* **2019**, *7*, 258. [CrossRef]
37. Crone, E.E.; Pelton, E.M.; Brown, L.M.; Thomas, C.C.; Schultz, C.B. Why Are Monarch Butterflies Declining in the West? Understanding the Importance of Multiple Correlated Drivers. *Ecol. Appl.* **2019**, *29*, e01975. [CrossRef]
38. Brower, L.P. Chemical Defence in Butterflies. In *The Biology of Butterflies*; Cornell University: Ithaca, NY, USA, 2010; pp. 109–134.
39. Brower, L.P.; Fink, L.S.; Walford, P. Fueling the Fall Migration of the Monarch Butterfly. *Integr. Comp. Biol.* **2006**, *46*, 1123–1142. [CrossRef]
40. Lukens, L.; Kasten, K.; Stenoien, C.; Cariveau, A.; Caldwell, W.; Oberhauser, K. Monarch Habitat in Conservation Grasslands. *Front. Ecol. Evol.* **2020**, *8*, 13. [CrossRef]
41. Cutting, B.T.; Tallamy, D.W. An Evaluation of Butterfly Gardens for Restoring Habitat for the Monarch Butterfly (Lepidoptera: Danaidae). *Environ. Entomol.* **2015**, *44*, 1328–1335. [CrossRef] [PubMed]
42. Baker, A.M.; Potter, D.A. Colonization and Usage of Eight Milkweed (Asclepias) Species by Monarch Butterflies and Bees in Urban Garden Settings. *J. Insect Conserv.* **2018**, *22*, 405–418. [CrossRef]
43. Morris, G.M.; Kline, C.; Morris, S.M. Status of Danaus Plexippus Population in Arizona. *lepi* **2015**, *69*, 91–107. [CrossRef]
44. Pegram, K.V.; Melkonoff, N.A. Assessing Preference and Survival of Danaus Plexippus on Two Western Species of Asclepias. *J. Insect Conserv.* **2020**, *24*, 287–295. [CrossRef]
45. Knapp, W.M.; Frances, A.; Noss, R.; Naczi, R.F.C.; Weakley, A.; Gann, G.D.; Baldwin, B.G.; Miller, J.; McIntyre, P.; Mishler, B.D.; et al. Vascular Plant Extinction in the Continental United States and Canada. *Conserv. Biol.* **2021**, *35*, 360–368. [CrossRef]

46. Plant Conservation Alliance. *National Seed Strategy for Rehabilitation and Restoration: Making Progress 2018*; Plant Conservation Alliance: Washington, DC, USA, 2018.
47. Whigham, D.F. Conserving Our Native Orchid Heritage—The What, How and WHen Behind the North American Orchid Conservation Center. *Nativ. Orchid. Conf. J.* **2012**, *9*, 24–31.
48. Coleman, R.A. The Wild Orchids of Arizona and New Mexico. Available online: https://www.amazon.com/Wild-Orchids-Arizona-New-Mexico/dp/0801439507 (accessed on 22 July 2022).
49. Managing Semi-Arid Watersheds: Watershed Basics—Public Land Management and Land Ownership in Arizona. Available online: https://www.fs.fed.us/rm/boise/AWAE/labs/awae_flagstaff/watersheds/basics/management.html (accessed on 15 July 2022).
50. Viglianisi, F.M.; Sabella, G. Biodiversity, Environmental Education and Social Media. *Biodivers. J.* **2011**, *2*, 195–200.
51. Di Minin, E.; Tenkanen, H.; Toivonen, T. Prospects and Challenges for Social Media Data in Conservation Science. *Front. Environ. Sci.* **2015**, *3*, 63. [CrossRef]
52. Frigerio, D.; Richter, A.; Per, E.; Pruse, B.; Vohland, K. Citizen Science in the Natural Sciences. In *The Science of Citizen Science*; Springer: Berlin/Heidelberg, Germany, 2021; pp. 79–96. [CrossRef]
53. Rhodes, C.J. Pollinator Decline—An Ecological Calamity in the Making? *Sci. Prog.* **2018**, *101*, 121–160. [CrossRef]
54. Wagner, D.L.; Grames, E.M.; Forister, M.L.; Berenbaum, M.R.; Stopak, D. Insect Decline in the Anthropocene: Death by a Thousand Cuts. *Proc. Natl. Acad. Sci. USA* **2021**, *118*, e2023989118. [CrossRef]
55. Plant Conservation Alliance. *National Seed Strategy for Rehabilitation and Restoration: Progress Report 2015–2020*; Plant Conservation Alliance: Washington, DC, USA, 2015.
56. Desert Refuge: Monarchs and Milkweeds in Arizona. USA National Phenology Network. Available online: https://www.usanpn.org/nn/desertrefuge (accessed on 17 February 2022).
57. Monarch Joint Venture. *Monarch Conservation Implementation Plan 2021*; Monarch Joint Venture: St. Paul, MN, USA, 2021.
58. Arizona Monarch Collaborative. Available online: https://www.azmonarchcollaborative.com/ (accessed on 9 March 2022).
59. Cotten, T.; Devos, J.; Keleher, C.; Lehr, S.; Marcum, S.; Miner, K.; Newmark, J.; Palmeri, D.; Sallabanks, R.; Bustos, M.U. *Western Monarch Butterfly Conservation Plan*; Western Association of Fish and Wildlife Agencies: Boise, ID, USA, 2019.
60. Arizona Monarch Collaborative—Resources. Available online: https://www.azmonarchcollaborative.com/resources (accessed on 22 July 2022).
61. Swarts, N.; Dixon, K. *Conservation Methods for Terrestrial Orchids*; J. Ross Publishing: Plantation, FL, USA, 2017; ISBN 978-1-60427-123-2.
62. Taylor, M.F.J.; Suckling, K.F.; Rachlinski, J.J. The Effectiveness of the Endangered Species Act: A Quantitative Analysis. *BioScience* **2005**, *55*, 360–367. [CrossRef]
63. Goettsch, B.; Hilton-Taylor, C.; Cruz-Piñón, G.; Duffy, J.P.; Frances, A.; Hernández, H.M.; Inger, R.; Pollock, C.; Schipper, J.; Superina, M.; et al. High Proportion of Cactus Species Threatened with Extinction. *Nat. Plants* **2015**, *1*, 15142. [CrossRef] [PubMed]
64. The IUCN Red List of Threatened Species. Available online: https://www.iucnredlist.org/en (accessed on 15 July 2022).
65. Gruber, J.S. Key Principles of Community-Based Natural Resource Management: A Synthesis and Interpretation of Identified Effective Approaches for Managing the Commons. *Environ. Manag.* **2010**, *45*, 52–66. [CrossRef] [PubMed]
66. Gavin, M.C.; McCarter, J.; Berkes, F.; Mead, A.T.P.; Sterling, E.J.; Tang, R.; Turner, N.J. Effective Biodiversity Conservation Requires Dynamic, Pluralistic, Partnership-Based Approaches. *Sustainability* **2018**, *10*, 1846. [CrossRef]
67. Ansell, C.; Gash, A. Collaborative Governance in Theory and Practice. *J. Public Adm. Res. Theory* **2008**, *18*, 543–571. [CrossRef]
68. Lashley, J. Solving Urban Conservation Issues through Botanic Garden Functions. Master's Thesis, Clemson University, Clemson, SC, USA, 2012; p. 43.
69. Krishnan, S.; Novy, A. The Role of Botanic Gardens in the Twenty-First Century. *CAB Rev.* **2016**, *11*, 1–10. [CrossRef]
70. Balding, M.; Williams, K.J.H. Plant blindness and the implications for plant conservation. *Conserv. Biol.* **2016**, *30*, 1192–1199. [CrossRef]
71. Borders, B.; Lee-Mäder, E. *A Conservation Practitioner's Guide 2014*; The Xerces Society: Portland, OR, USA, 2014.
72. Hanson, N.; Ross-Davis, A.L.; Davis, A.S. Growth and Survival of Two Western Milkweed Species: Effects of Container Volume and Fertilizer Rate. *HortTechnology* **2017**, *27*, 482–489. [CrossRef]
73. Topping, M.L.; Dumroese, R.K.; Pinto, J.R. Successfully Storing Milkweed Taproots for Habitat Restoration. *NPJ* **2019**, *20*, 48–58. [CrossRef]
74. Steger, C.; Klein, J.A.; Reid, R.S.; Lavorel, S.; Tucker, C.; Hopping, K.A.; Marchant, R.; Teel, T.; Cuni-Sanchez, A.; Dorji, T.; et al. Science with Society: Evidence-Based Guidance for Best Practices in Environmental Transdisciplinary Work. *Glob. Environ. Change* **2021**, *68*, 102240. [CrossRef]
75. Fotiadis, A.; Polyzos, S.; Huan, T.-C.T.C. The Good, the Bad and the Ugly on COVID-19 Tourism Recovery. *Ann. Tour. Res.* **2021**, *87*, 103117. [CrossRef] [PubMed]
76. The ICBOs and Allies Workgroup Understanding the Impact of Equitable Collaborations between Science Institutions and Community-Based Organizations: Improving Science through Community-Led Research. *BioScience* **2022**, *72*, 585–600. [CrossRef] [PubMed]
77. *Missouri Botanical Garden: Consolidated Financial Statements and Supplementary Information 2014*; Mayer Hoffman McCann: St. Louis, MI, USA, 2014.

78. *The Atlanta Botanical Garden Inc.: Financial Statements 2019*; Smith & Howard: Atlanta, GA, USA, 2019.
79. *Desert Botanical Garden Inc. and Affiliates: Consolidated Financial Statements 2020*; Desert Botanical Garden: Phoenix, AZ, USA, 2020.
80. Wright, A.J.; Veríssimo, D.; Pilfold, K.; Parsons, E.C.M.; Ventre, K.; Cousins, J.; Jefferson, R.; Koldewey, H.; Llewellyn, F.; McKinley, E. Competitive Outreach in the 21st Century: Why We Need Conservation Marketing. *Ocean. Coast. Manag.* **2015**, *115*, 41–48. [CrossRef]

Journal of
*Zoological and
Botanical Gardens*

MDPI

Article

Local Plant and Insect Conservation Evaluated with Organizational Identity Theory

Lily Maynard *, Bailey Cadena , T'Noya Thompson, Valerie Pence, Megan Philpott , Mollie O'Neil, Mandy Pritchard, Julia Glenn, Bridget Reilly, Jordan Hubrich and David Jenike

Cincinnati Zoo & Botanical Garden, Cincinnati, OH 45220, USA
* Correspondence: lily.maynard@cincinnatizoo.org

Abstract: With a range of programs focused on local plant and insect conservation, the Cincinnati Zoo & Botanical Garden works with partners and our communities to restore landscapes and thriving ecosystems for wildlife and people. We used organizational identity theory (OIT) to evaluate the current strategies and practices of five programs and determine opportunities for adaptation to better achieve our organizational conservation goals. Case studies ranged from habitat restoration of wetlands and gardens to community engagement to encourage individual gardeners and the reintroduction of endangered plants and insects. We present program characteristics and how collaborative partnerships facilitate opportunities for zoos to lead the conservation of local flora and fauna. The OIT framework reveals components critical to strategy implementation and best practices relevant to other zoos, aquariums, and botanical gardens when evaluating their collaborative conservation initiatives.

Keywords: zoos; local wildlife; conservation; collaborative partnership; habitat restoration; pollinator garden; community engagement; reintroduction

Citation: Maynard, L.; Cadena, B.; Thompson, T.; Pence, V.; Philpott, M.; O'Neil, M.; Pritchard, M.; Glenn, J.; Reilly, B.; Hubrich, J.; et al. Local Plant and Insect Conservation Evaluated with Organizational Identity Theory. *J. Zool. Bot. Gard.* **2023**, *4*, 214–230. https://doi.org/10.3390/jzbg4010019

Academic Editors: Ursula Bechert and Debra C. Colodner

Received: 30 September 2022
Revised: 4 February 2023
Accepted: 20 February 2023
Published: 3 March 2023

1. Introduction

Human changes to landscapes are driving the biodiversity crisis for plants and animals and the decline of functional ecosystems [1–4]. Biodiversity degradation is more concentrated in urban areas with higher intensities of human modifications to the landscapes [5]. However, people can also provide solutions; communities can be mobilized to reduce threats and support conservation [6].

Insects and plants are sensitive to human-caused stressors and ecosystem changes [7–9]. In the last 30 years, terrestrial insect populations have declined by 24% [4]. While many people are underinformed or unaware of the complex issue of declining insect populations [10], we need people to identify, implement, and embrace interventions that support insect populations, which are critical for ecosystem health, services, and resilience [11–13]. Plants are also at great risk of biodiversity loss, with an estimated 40% of plant species at risk of extinction due to threats such as climate change, habitat loss, and invasive species [14]. The tendency for organizations to focus on animal conservation can result in plants being ignored and their conservation underfunded, despite having the most species on the endangered species list [15,16]. Many plant species have dispersal mechanisms that make them unable to adapt to quickly changing environmental conditions and threats, and as a result both in situ and ex situ conservation methods are often recommended [17]. Zoos, aquariums, and botanical gardens can lead their communities to embrace and implement plant and insect conservation efforts. Conservation research has historically focused on larger, intact natural areas, yet scientists have identified urban and suburban landscapes as settings for biodiversity conservation through landscape interventions, such as diverse plant installations and gardens [18]. Residential lawns engulf almost 40% of the total area of the United States [19], yet these monocultures or "green deserts" provide little food or habitat for insects or other wildlife [20]. Only 18% of a typical yard includes more diverse plants, trees, and shrubs that could provide habitat or resources [21].

Plant restoration can create welcoming landscapes for wildlife [22–24]. For example, diverse plants in urban gardens have a strong, positive influence on insect populations [25–29]. Insect conservation programs recommend a diversity of native plant species to increase viable landscapes [4,30–33]. Community members in residential areas, no matter if they have a sprawling area in their yard or just a few flowerpots, can have a positive impact on insect populations [34]. With more diverse green spaces and corridors, cities can provide vital refuges for wildlife [35].

1.1. Collaborating with Communities

Conservation issues are complex. From economic and political decision-makers to the ecological drivers, social conditions, and characteristics of the communities, the factors that influence environmental conservation are diverse, nuanced, and often connected to each other (e.g., PESTLE analysis in Maynard et al., 2022 [36]). However, engaging diverse communities and organizations can support impactful change with more perspectives present and empowered to share creative solutions [37]. These challenges have shifted how organizations use community-based approaches to create sustainable actions through diverse partnerships and networks [38]. We predict that more organizations, individuals, and communities involved in conservation efforts will lead to more potential for conservation impact through resource-sharing and coordination [39,40].

Furthermore, zoos and aquariums can work together to lead their millions of visitors and nearby communities to participate in conservation [40]. In recent years, zoos have invested more active staff involvement in conservation projects to step into this leadership role [41]. While zoos have historically branded themselves separately for individual conservation projects [42], complex conservation issues exceed one organization's reach, so collaborative partnerships are vital for success [40].

To combat the biodiversity crisis, conservation organizations must improve leadership strategies to better serve communities and meet organizational goals [43]. Vision, values, and partnership-building are critical components in conservation leadership and the resources invested to influence the environment [44]. This study investigates the unclear relationship between organizational leadership and the success of conservation goals [45,46].

Conservation initiatives need community support to combat complex challenges [47]. Over the past few decades, conservation organizations have focused more on localized, community-based work [48,49]. When organizations work, serve, and support communities, a positive and transferable leadership presence can be cultivated to reinforce their reputation [50]. Community involvement can amplify awareness of environmental issues while supporting research project needs [51].

Adaptive management based on scientific evaluation is critical for conservation efforts within dynamic communities and landscapes [52]. Community Science (CS) promotes efficient data collection for program evaluation alongside local audience engagement [53]. CS allows people to be intricately involved in science through organized mobilization to collect data and support research projects [54]. CS programs need collaborative engagement, organized structure, and relevance to be considered successful [55].

When done well, CS programs can provide high volumes of accessible data collection through volunteer participation. The annual estimate of CS participants ranges from 1.3–2.3 million volunteers who donate up to USD 2.5 billion in kind [56]. The growth within these community programs and a stronger focus on leadership strategies [44,52] highlight the need for conservation programs to evaluate their effective connections to their communities and regional landscapes [57].

1.2. Conservation Identity Evaluation

Conservation strategies require evaluation and adaptive management to ensure their goals are being achieved [58]. We conducted this study to evaluate the current activities and impact of the Cincinnati Zoo & Botanical Garden's (CZBG) local programs to restore landscapes for plants and insects while inspiring community involvement. We assessed five programs that include different species and activities to examine our progress toward our local wildlife coexistence goals.

Organizational identity theory (OIT) [59] provided the framework for the evaluation in this study. OIT outlines components of an organization's strategic plans across six constructs to enable constructive evaluation. From the foundations of identities and desired outcomes to the application of intentions into active projects and invitations for stakeholders to join in, the six constructs highlight connections between abstract aspirations and their application on the ground [59]. (1) The central construct defines the core features and missions of an organization. (2) The enduring construct explores how long an identity has endured, and even strengthened, over time to understand how an organization has prioritized this identity over other possible strategies [59]. (3) The distinctive construct highlights what makes an organization and its activities unique. (4) The goal-oriented construct assesses whether the organization set deliberate intentions as important precursors to organization's decision-making for future activities [60]. (5) The responsibility construct examines how an organization creates legitimacy when it influences its stakeholders toward acting in support of its mission [60]. (6) The performance construct facilitates the formative evaluation of current performance against the organization's goals to promote reflection on opportunities for improvement [60].

OIT is similar to a logic model framework of inputs, outputs, and outcomes, yet OIT aids the evaluation to dig deeper by adding factors of the strategic intentions of the organization [61]. By including leadership's perspectives and reflections on goals and calls-to-action, OIT enables a thorough evaluation across the organization's initiatives with many perspectives considered [60].

Organizations seeking to promote a cause and mobilize their community can use OIT to design and monitor their strategies [62]. Organizations are more likely to achieve their goals when they reflect on and manage their identities [63]. The intentional activities and impact desired by organizations are informative components of their identities, and through reflective evaluation new opportunities can be revealed to increase their effectiveness.

Zoos and aquariums historically participated in conservation by distributing revenue to other organizations, yet their staff could feel disconnected from conservation [6]. Creating opportunities for staff to be directly involved in a conservation project beyond the walls of the organization's grounds allows diverse individuals throughout the organization to connect with the identity by participating in the larger community and landscape. Looking outside of the organization, the formation of collaborative partnerships on the projects extends the identity to other organizations throughout the community for additional resources and individuals to join in and expand the project's reach. These are key components of the distinctive construct for OIT.

Applying the OIT framework can aid zoological institutions and botanical gardens to compare their goals against their activities and progress. The OIT framework organized our review of CZBG's strategies and activities to document both successes and opportunities for conservation identities to be strengthened [64,65]. OIT helps to assess the relationships between organizational intentions and the characteristics of identities in practice [59]. We invited project leaders across the five zoo case studies to participate in the collaborative evaluation, as OIT predicts that leaders of an organization and project team know and use their organizational identity to guide project decisions toward their goals [60]. If zoos strive to lead communities to participate in wildlife conservation [66], then this assessment of the local plant and insect conservation programs at CZBG will provide relevant example methods of formative evaluation by using the OIT framework.

1.3. Project Goals

Conservation strategies for local species, such as insects and plants, can be led by nonprofit organizations with proactive and extensive collaborations and communications throughout their communities [67]. This study builds on similar evaluations of collaborative conservation initiatives (e.g., Maynard et al., 2022 [36]), which assessed whether organizational activities both connect directly to the threats species face and mobilize more people to take action to increase efficient resource-sharing, thereby creating a movement beyond the reach of one organization. Organizational identity theory provides the framework for evaluating the local conservation programs at CZBG.

2. Methods

2.1. Project Site

CZBG is located in the city of Cincinnati, OH, USA. With a population of approximately 300,000 within the city and 2 million people in the region [68], approximately 1.7 million people visit CZBG annually. Urban landscapes surround CZBG, but we engage our visitors in conservation activities to restore diversity through native plants and pollinator habitats. CZBG has several programs to promote local plant and insect conservation, with the goal of facilitating increased biodiversity and vibrant landscapes and mobilizing our audiences. This study will contrast our local conservation activities against our organizational identity and goals to assess effectiveness.

Understanding the conservation impact strategic plan for the organization provides critical context before jumping into evaluating the local conservation case studies. Coexistence between wildlife and people is the foundation of the CZBG conservation impact strategy. The four coexistence goals of (1) impact, (2) mobilize, (3) elevate, and (4) justice are mechanisms for measuring progress for all projects implemented under this strategy. The research team used these four coexistence goals to contextualize and interpret identity evaluation using OIT constructs.

The organizational conservation impact strategy at CZBG includes the following statements:

The goal of the CZBG conservation impact strategy is to activate our staff, visitors, stakeholders, and audiences to be inspired by wildlife to actively participate in conservation. Our Conservation Impact initiatives are unique, with our focus on coexistence and proactive calls-to-action for every project and every audience member.

Through an empowering alliance promoting benefits for all, we can collaborate, advance science toward impact, enhance solutions and reduce conflict, and sustain healthy landscapes and resources for wildlife and people. By leveraging the zoo and botanical garden's deep expertise and celebrating our unique animals and plants, we are building a brighter future for wildlife and people.

Our coexistence goals are:

1. Impact—Increase our direct impact on threats to wildlife and field partners' needs, with zoo staff on the ground co-developing projects with our partners.
2. Mobilize—Increase our mobilization of zoo staff, visitors, and local communities in strategic action for sustainable resource use and wildlife coexistence.
3. Elevate—Increase our team's expertise, external presence, reputation, and credibility by training future conservation leaders and leading collaborations.
4. Justice—Increase our active inclusion, equity and justice efforts by investing in and supporting diverse, indigenous experts with proactive partnerships.

2.2. Screening Criteria

To evaluate whether the organizational conservation identity of CZBG is developed in our local conservation projects, we assessed five case studies of local plant and insect conservation projects using the OIT framework (Figure 1). The case studies were selected from a complete census of CZBG's local conservation projects. Screening criteria required the projects to have zoo staff directly involved in the projects' activities rather than zoo funds being used by other partner organizations. All five local conservation programs revealed in the census were selected for the study as they met the screening criteria.

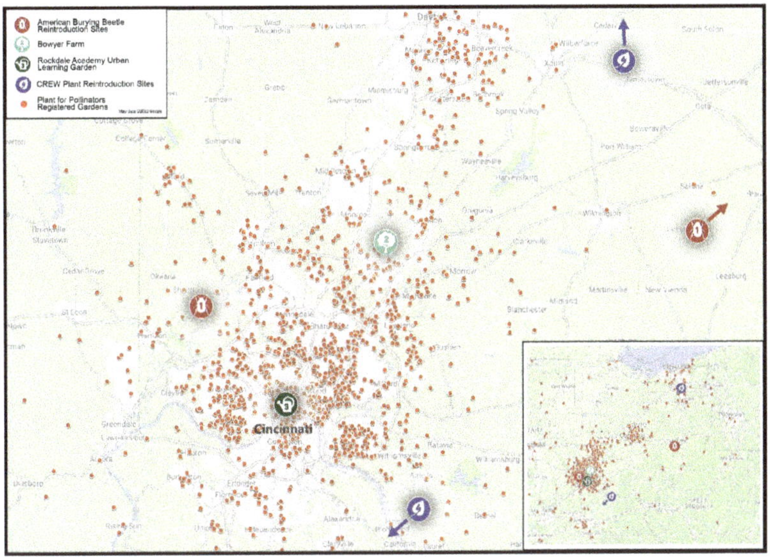

Figure 1. A map depicting the approximate location of the five CZBG local conservation case studies across Ohio, as well as parts of Kentucky and Indiana. These widespread programs create an abundance of vibrant landscapes by mobilizing community members in CS, allowing for greater outcomes from staff engagement. Program icons are defined in the top left key. The bottom-right square depicts the zoomed-out scope of projects across Ohio and the surrounding states. The reintroduction sites for (1) ABB and (4) CREW are approximately marked with an arrow to indicate direction without identifying specific locations of the highly sensitive, endangered species.

2.3. *Organizational Identity Theory Constructs*

We used OIT as the deductive framework for our evaluation. We explored the features of the projects as potential evidence of CZBG's conservation identity by sorting details into the six constructs in the OIT framework for each case study (Figure 2). OIT has three core constructs that describe the projects' (1) central, (2) enduring, and (3) distinctive characteristics as they establish legitimacy and recognizability [2]. OIT then includes three constructs to assess the identity's application attributes and success in practice, which are (4) goal statements, (5) activations of responsibility, and (6) perceptions of current performance [69,70]. By categorizing the components of OIT, the constructs organized our qualitative content analysis into a formative evaluation of the strategic goals of our organization.

Central: The central construct of what is core to the conservation efforts is the foundation for an organizations' identity and strategies [61]. Evaluating an organizations' programs and initiatives can reveal their potential impact represented by their active identity, which is more informative than reviewing their mission statements and brands [6]. The evaluation of each case study separately will use the central construct to uncover the core characteristics of the programs.

Enduring: The programs' enduring identity is impacted by their age and the growth of their activities over time.

Distinctive: An organization makes itself distinctive by emphasizing its identity with calls to action and exemplary messages to both internal staff and external stakeholders [71].

Goal-oriented: The goal-oriented construct links an organization's core missions with the strategies they later implement by defining their desired outcomes [64].

Responsibility: Using this responsibility to promote new participation in activities, an organization progresses toward its goals and amplifies more resources for the project [64].

Identity-performance: This construct is measured by reviewing existing outputs and metrics for desired outcomes to assess current trajectory and opportunities for intentional adaptation.

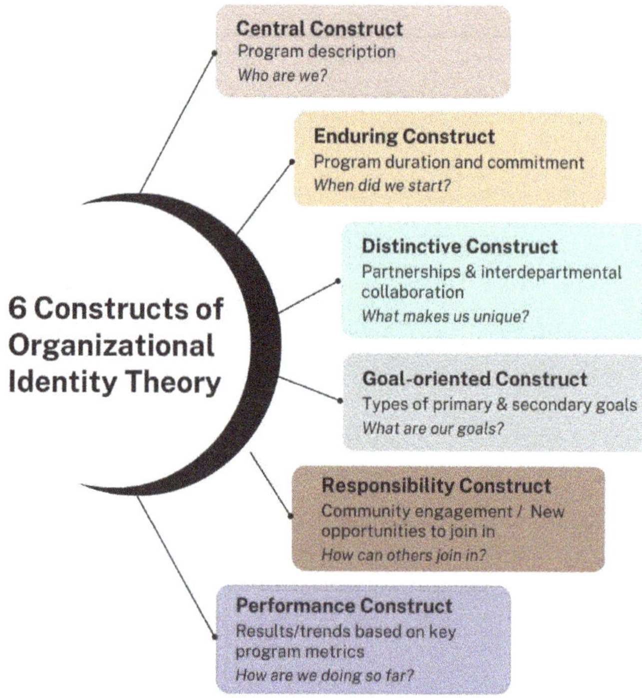

Figure 2. Descriptions and important components of the six constructs of organizational identity theory.

2.4. Data Collection and Analysis

We conducted a collaborative content analysis with project partners from the five case studies to review the progress reports, project management documents, activities, and metrics for each project. The research team met with the case study partners to document details about each program. OIT is an internal evaluation tool that uses inward-looking criteria to evaluate program effectiveness and reveal opportunities to strengthen organizational identity and program alignment. This is why the partners who were interviewed did not include anyone outside the organization.

The research team compiled details about each case study and categorized the materials into tables for the six constructs from OIT. We used an iterative process of reviewing and discussing the data categorizations to reach consensus from the research team and collaboratively incorporated perspectives from the projects' partners [72].

We used purposive sampling to identify project partners as participants for the case study reviews (n = 10). Screening criteria for project partners to participate in the collaborative content analysis discussions included having worked on that project for at least one year to ensure their understanding of the project details, their availability for review meetings, and their willingness to discuss their perspectives and project activities. The research team acknowledged the potential for bias toward positive comments from participants due to their involvement in the projects and their investment in its success; however, we framed the evaluation around the value of identifying opportunities for improvement to encourage participants' honesty.

This collaborative content analysis evaluated each project's features across the six constructs of their organizational conservation identities. We discussed these details and assessed their level of activation to achieve this piece of the identity development. We also highlighted any less developed constructs as opportunities for enhanced identity development to achieve greater potential impact.

The research team initially tried using an online survey and email inquiries to collect details about the projects, yet the simplified questions did not elicit rich qualitative details about the programs. After adapting to in-person meetings and co-reviewing documents and reports, the collaborative process of including participants in collecting the program details helped the research team to document the projects during the evaluation process.

3. Case Studies

In the following section, we evaluate each local conservation program by highlighting its relevant components within the six organizational identity constructs with illustrative details and examples. The three core constructs present the foundation of the programs, while the three applied constructs reveal their identity in practice [64]. The case studies had some well-developed identities and some with incomplete components, emphasizing different features and leadership decisions. The less established constructs reveal opportunities for adaptive management and growth for these case studies to better achieve their local conservation goals.

Organizations' leaders represent and guide the organization's identity into its tangible strategies and projects [73]. By conducting a collaborative content analysis with the leaders (n = 10) of five local conservation programs at the Cincinnati Zoo & Botanical Garden, we had the most informed participants share essential details regarding their current activities, their goals, and opportunities for improvement.

3.1. American Burying Beetle (ABB)

Central: This project involves captive breeding, reintroductions, and presence/absence surveys in Ohio of the ABB. The recovery of this state endangered and federally threatened species relies upon collaboration between multiple organizations and government agencies. Each summer, founders are collected by CZBG World of the Insect Keepers from the sandhills of Nebraska with the help of Nebraska Game and Parks Commission, the USFWS, and The Wilds and brought to Ohio. Those beetles and their offspring are bred at CZBG and The Wilds to produce a large release population for the following summer. Reintroductions have been held at The Fernald Preserve with cooperation from the Department of Energy and at The Wilds, which is partially supported through a grant with the Ohio Department of Natural resources and in-kind support from the CZBG.

Enduring: The ABB program started over 30 years ago, with CZBG participating in ex situ breeding of the endangered species. The program's identity has endured and evolved with the additional active participation in local reintroduction efforts starting in 2011.

Distinctive: This project involves diverse staff in the field activities implementing the interventions and monitoring the target species, such as when zoo staff create field equipment for the ABB releases. Collaborations with external partners drives the program forward with regular communications between organizations, shared project management, coordination of activities, and collaborating on fundraising. The ABB reintroduction team noted that open knowledge sharing of best practices between partners involved in breeding and reintroduction efforts has been vital to the success of this initiative.

Goal-Oriented: The goal of this project is to support the recovery of this state endangered species through collaboration between multiple organizations and government agencies for beetle breeding in CZBG care and reintroduction efforts locally. When reflecting these project goals against the four coexistence goals for the CZBG strategic plan, the case studies prioritized different facets of the coexistence goals for documenting progress (Table 1). This project was built around the priority of impact for specific species and habitats. Other coexistence goals could be considered for future growth of the program.

Table 1. Program details for the five case studies representing the enduring, distinctive, goal, and responsibility constructs from OIT.

Enduring Construct		Distinctive Construct		Goal Construct		Responsibility Construct	
CZBG Program	**Age of the Project (Years)**	**Types of Partners**	**Inter-Departmental Collaboration Across CZBG Outside of the Project Team?**	**Priority Coexistence Goal**	**Secondary Coexistence Goal**	**Call-to-Action for Community Involvement?**	**Number of Partner Organizations**
American Burying Beetle (ABB)	31	Federal government agencies, state fish and wildlife agencies, other zoos, state and local parks	Yes	Impact	Elevate	No	8
Bowyer Farm	10	Federal government agencies, scientists and universities, extension agents, plant nurseries, restoration specialists and contractors	Yes	Impact	Mobilize	Yes	10
CREW plant reintro-ductions	10	Federal government agencies, scientists and universities, state and local parks	No	Impact	Elevate	No	10
Plant for Pollinators (P4P)	3	Individual participants, sponsors and donors	Yes	Mobilize	Impact	Yes	19
Rockdale Urban Learning Garden (ULG)	2	City public schools, local community organizations, sponsors and donors	Yes	Justice	Impact	Yes	5

Responsibility: The calls-to-action for zoo audiences vary in strength across the projects (Table 1). The ABB program does not communicate a call-to-action for participation in the program. The reintroduction-focused projects are handling endangered species and do not provide a call-to-action for the public to get involved. This may be due to the endangered species being rare, so community members would rarely see them to participate in community science. The programs are described and celebrated on the zoo's social media to raise awareness and indirect support for conservation action, but they do not convey direct ways for zoo audiences to get involved in protecting these plants and insects. However, communications could improve with messages beyond awareness raising about ABB to include ways to create habitat for other native insects important for our ecosystems.

Performance: 1833 ABB released into the wild from CZBG population since 2013. Twenty-four beetles recaptured to check survivability, 20 offspring captured to check the next generation's success, and an estimated 2630 larvae produced in the wild. Overall, the ABB species was down-listed federally so progress is being made, though it is still listed as endangered in the state of Ohio.

3.2. *Bowyer Farm*

Central: Bowyer Farm is over 600 acres (243 hectares) and supports a variety of projects including wetland reclamation started in partnership with the US Department of Agriculture. Initial habitat restoration focused on rebuilding a vernal pool from altered soy and corn fields to create habitat for amphibians, and over time additional habitat features have been restored for other wildlife species. This farm was willed to the CZBG under the guideline that it could not be developed unless doing so would further the mission of CZBG. The farm also grows native plants for public sale and planting in CZBG pollinator gardens, as well as browse growth for zoo animals and animal enrichment. Bowyer is also the location for an upcoming solar panel installation (size = 25 MW) that will help CZBG move from net-zero to net-positive energy usage.

Enduring: The older programs have endured and grown from their original goals to add additional activities to the sites and new species to the conservation management efforts. The conservation program at Bowyer Farm started with a focus on recreating a vernal pool for amphibians, but over time additional habitats have been restored to provide year-round water for other species, including plants, waterfowl, migratory birds, and other amphibians. This program has endured and grown over time, enhancing CZBG's commitment to the conservation identity.

Distinctive: This project involves diverse staff in the field activities of habitat restoration and plant nursery maintenance. For example, diverse zoo staff join wildlife observation events at Bowyer Farm to monitor bird, insect, amphibian, and reptile species in the restored landscapes.

Goal-Oriented: The goal of the project is to restore habitats from farmland back into wetlands, forests, and prairie and manage the landscape for as much diversity in plants, birds, amphibians, and reptiles as possible. Later on, additional goals were established, such as creating a native plant nursery, hosting events, and welcoming external partners to conduct research. These goals start with a focus on the impact for specific species and the habitats restored across the landscape, with additional coexistence goals acknowledged around ways to mobilize audiences at the farm and elevate the zoo as a conservation leader and host in the space. However, Bowyer Farm has an opportunity to grow into the untouched goal of how to promote environmental justice in nearby underserved communities.

Responsibility: Bowyer Farm offers some volunteer events to support habitat restoration activities, but it is not open to the public every day. However, on the Saturdays when the farm is open, an average of two thousand people come to purchase native plants and learn from the horticulture team.

Performance: 205 bird species, 13 amphibian species, and 7 reptile species have been observed onsite at the Bowyer Farm, demonstrating the viable habitat welcoming diverse species from the prior corn and soy farm fields. Additionally, the CZBG horticulture team have grown over 300 native plant species and sold 12,522 individual native plants, documenting the extent of increased supply of native plants provided by the Bowyer Farm.

3.3. CREW Plants

Central: This project focuses on producing plants for habitat restorations through in situ reintroduction and preserving plants in long-term storage so they can be used in future restorations. Scientists in the Plant Research Division at the Lindner Center for Conservation and Research of Endangered Wildlife's (CREW) form critical partnerships with other leading conservationists and governmental and non-governmental organizations to achieve CREW's mission to Save Species with Science. CREW plant scientists are actively involved in local conservation for endangered species. In partnership with the US Fish & Wildlife Service, other federal and state government agencies, parks, and nonprofit organizations, the CREW team propagates endangered plants for reintroduction to restore the biodiversity of landscapes (Figure 1). For example, plants propagated at CREW for a federally endangered sandwort in the project scope have been sent back to managers with the Daniel Boone National Forest for use in restorations [74].

Enduring: This program was created 10 years ago. The CREW endangered plant research and reintroductions have added new species over time, including species from across the United States and rare Hawaiian species, further solidifying this organizational identity with this long-term commitment.

Distinctive: Unique external partnerships with the government agencies and other partners are the driving force behind this important program for endangered species. However, interdepartmental collaboration outside of the project team was present for all the case studies, except for the CREW endangered plant reintroductions (Table 1). This specific program focuses on staff propagation of the plants, but CZBG staff do not lead the reintroduction activities and so cannot invite others to participate.

Goal-Oriented: The goal of the CREW endangered plant program is to reproduce and distribute target species to partners for reintroduction and habitat restoration efforts. This prioritizes the impact coexistence goal for the specific species, and the active support of government agencies and partners supports the coexistence goal of elevating the zoo as a conservation leader.

Responsibility: The CREW endangered plant research does not communicate a call-to-action for zoo visitors, community members, or other staff to get involved in the program. Due to the endangered species, direct participation is restricted. However, this program can enhance its proactive communication using these species as inspiration for plant conservation and promoting diverse plants in our landscapes.

Performance: Nine plant species were reproduced in the CREW plant lab, and over 1400 individual plants were reintroduced by being planted in the wild.

3.4. Plant for Pollinators

Central: Plant for Pollinators (P4P) is a public engagement and action campaign to increase pollinator habitat in the Greater Cincinnati region and beyond by (1) connecting both new and experienced gardeners to resources regarding planting for pollinators, and (2) encouraging people to register their new or existing pollinator gardens through the P4P challenge so that we can track the expansion of pollinator habitat across the region and country. The P4P program is a cross-departmental partnership with input from our Administrative, Horticulture, Insect, Conservation, and Education departments. P4P is also cross promoted with the Zoo's annual native plant sales, which seek to raise money for the Zoo while also increasing native plant habitat across the tristate region. P4P is sponsored by Simple Truth and supported by over 20 local partners, whose funds enable the creation of promotional materials and overall sustainability of the program.

Enduring: This newer project began in 2019. It reinforces CZBG's previous commitment to plant and insect conservation by promoting our knowledge from our accredited botanical garden but needs more time to display its influence on the zoo's identity as it strengthens and evolves over time.

Distinctive: Diverse staff are involved in the gardening kit development, promotions, and gardening events for P4P. Staff and external partners are also invited to get involved and register their gardens with the P4P challenge. With a long list of partners and sponsors, this program has created a network of collaborators to promote creating pollinator habitat in the region.

Goal-Oriented: The goal of this project is to create more pollinator-friendly habitats and increase community involvement to mobilize individuals to plant diverse plants for pollinators and register their gardens. The primary coexistence goal prioritized by this program focuses on mobilizing communities to join the P4P program, and subsequently addresses impact for species and habitats. Additional goals for leadership elevation and environmental justice by supporting underserved communities are opportunities for future growth of this program.

Responsibility: Plant for Pollinators has the most accessible public participation opportunity of the case studies, because inviting any individual to plant and register their gardens is the focus of the program. The mobilization of individuals to coordinate efforts ensures a network of gardens is created in urban habitats and subsequently addresses the impact goals of habitat creation for pollinators.

Performance: Over 2979 individuals' gardens were registered in the challenge (Figure 1), over 70 pollinator kits were sold for individuals' gardens, and over 3000 people were reached in pollinator-focused horticulture talks.

3.5. Rockdale Urban Learning Garden

Central: Before it was transformed into a biodiverse working garden and living classroom, the Urban Learning Garden (ULG) at Rockdale Academy was an unused, one-acre field of turfgrass. As part of the 2021 Community Makeover led by CZBG, the Reds Community Fund, P&G, and Cincinnati Children's Hospital, this homogenous landscape was transformed into a one-of-a-kind learning space. The garden now hosts over 10,000 plants, made up of perennials, shrubs, fruit trees, vegetables, and herbs, as well as a greenhouse powered by solar panels that were installed on the elementary school's roof. The zoo has a full-time horticulturalist maintaining the garden with the help of volunteers, and occasionally students, who also use the space for community and after-school events. The garden is a unique example of long-term socioecological investment in that the zoo has committed to, through labor and educational programming, ensure that the garden does not fall into a familiar pattern of beautification projects that later suffer from neglect. While the community engagement components of the garden are still being explored, its installation and subsequent year of care have undoubtedly changed the environmental makeup of the urban neighborhood of Avondale by attracting more wildlife, especially a wide array of pollinators.

Enduring: This program began in 2020 and has yet to evolve into new dimensions of conservation identities for CZBG. However, since the ULG was built, many partners have visited and expressed their interest in the program; this highlights opportunities for reinforcing this conservation identity's importance as new partners join in and built onto the zoo's project.

Distinctive: Internally, diverse zoo staff have been involved in building the garden and hosting community events there. Externally, many large corporations partnered to create this unique space for a deserving underserved school and community. These diverse partners reinforce the distinctive conservation identity created by the ULG.

Goal-Oriented: The goal of this project is to nourish, educate, and inspire the children of Avondale in Cincinnati, OH through horticulture, science, and the outdoors. As such, the primary coexistence goal that drove this program's creation was to facilitate justice for the community and school through the ULG. This leadership is then followed secondarily by the pollinator habitat created for the impact goal, the elevation of the zoo with many partners, and the opportunities to mobilize the target community in the novel space.

Responsibility: The Rockdale ULG program mobilizes a smaller scope of the local community than the P4P program by focusing on residents around the zoo in the Avondale neighborhood with events and opportunities in the community garden. However, with diverse events and educational experiences, this case study provides many different calls-to-action for people to join in.

Performance: Over 350 Pre-Kindergarten to 6th grade students and their families in the community and 60 staff from Rockdale Academy were reached with events in the Urban Learning Garden.

4. Discussion

The methods used in this evaluation were useful, informative, and provided opportunities for relationship-building. The collaborative content analysis generated detailed information regarding the case studies. Purposive sampling of project leaders and partners was well-received by the participants and led to abundant and valuable information regarding the programs. By documenting more components of each program, new details regarding their successes and growth were shared across the zoo, which created new relationships and opportunities to celebrate or even envision new collaborations.

The project team faced some challenges when gathering the diverse data regarding the wide-ranging projects. Each team and program are organized differently, so they needed to be met with in-person to gather the needed information. These conversations led to co-developing ideas for growth opportunities for the program, an important outcome created in partnership with the program leaders to facilitate their willingness to adapt their programs based on the evaluation findings. While the process took longer than an online survey with more meetings and participants involved, the outcomes were richer for the evaluation and the participatory process enabled the program leaders to prepare to use the results. Direct communication and an emphasis on identifying opportunities for improvement encouraged participants to highlight not only the program successes.

The qualitative exploration of the projects' characteristics revealed both CZBG's local conservation programs' strengths and gaps for future growth. The project leaders shared how the case studies reinforce unique conservation identities at CZBG. Each team strongly identified with their program, its unique opportunities, and pride in their conservation efforts. Furthermore, the central characteristics of the five case studies revealed foundational components and vivid details regarding the projects' development with collaborative partnerships and intentions to facilitate habitat restoration and community engagement. The five programs differ in their scope (Figure 1), but each has a clear definition of its target area and species. The case studies demonstrate CZBG commitment to partnerships and active interventions for native insects and plants. The differing geographical ranges and types of partners highlight multiple strategies for impact that influenced the creation of the programs.

4.1. Partnerships

The CZBG local conservation programs all have detailed internal and external collaborations that highlight their strong, distinctive identities. In particular, collaborations with external partners drive the five programs forward. Active partnerships across the case studies include regular communications between organizations, shared project management, coordination of activities, and collaborating on fundraising. The types of organizations that CZBG partners with to make these projects possible include federal government agencies, state fish and wildlife agencies, universities and scientists, other zoos and aquariums, other nonprofit organizations, public school districts, state and local parks, and activity specialists such as plant nurseries and habitat restoration experts (Table 1). The screening criteria for the case studies included in this study required some active zoo staff involvement in the programs, so all five programs achieve this minimum level of internal involvement. We chose to emphasize this baseline of active partnership with staff commitment to the projects because the majority of conservation programs in zoos and aquariums do not achieve this minimum, and instead send funds to other organizations with minor staff involvement [6].

Multi-organization collaboration makes these complex conservation projects possible, no matter if they are reintroducing endangered beetles [75] or inspiring pollinator habitat restoration and community science [36]. These projects are possible due to the wide-ranging partnerships and active communications and resource-sharing between stakeholders, from the government funding of restoration at the Bowyer Farm to the many sponsors and donors for the Plant for Pollinators and Rockdale ULG.

CZBG can reflect on the types of partners effective in one program as potential stakeholders to strengthen the other programs. For example, government agencies or other non-profit organizations interested in pollinator habitat restoration could better enhance the P4P project beyond the individual community members and donors currently participating. Additionally, the public school partners unique to the ULG might find partnering around the four other projects valuable to their educational and community engagement goals. These four projects can reflect on the learnings from the ULG team to better include educational and community partners in their activities in the future.

4.2. Coexistence Goal Assessments

These local conservation programs are driving progress toward CZBG's coexistence goals of impact, mobilize, elevate and justice. Though many of the local conservation projects at CZBG connect to each of the four coexistence goals for progress in our strategic plan, the projects' priorities were imbalanced. The older case studies focused on the impact of endangered species or habitats first. For example, the ABB and CREW projects could improve by considering how to mobilize more community and audience members to extend their reach and resources. Similarly, the Bowyer Farm project could establish new opportunities to empower nearby underserved communities and promote environmental justice through inclusion. The reordering of the priorities in the design of the newer projects to emphasize community mobilization and facilitating justice reflects the evolution of a multi-faceted definition of conservation success, and therefore can serve as a model for updating older programs. On the other hand, the newer projects of P4P and ULG while well-known locally, could better elevate CZBG by highlighting our organizational identity around conservation with intentional storytelling regarding these projects. Reflecting these coexistence goals across the projects and integrating learnings across the departments will amplify a clearer and impressive conservation identity for CZBG.

The OIT constructs also highlighted some opportunities for thew potential growth of the programs to better achieve their conservation goals. The reintroduction programs were distinctive in their absence of opportunities for other zoo staff to participate, nor did they have a clear pathway for inspired zoo audiences to join in. This highlights an area for CZBG to consider growing their programs to achieve even greater impact with additional people involved from our large audiences of zoo visitors and social media followers. Two of the case studies also lacked calls-to-action for public audience involvement. These local conservation programs are making progress toward supporting their organizational identities, but could improve by acknowledging opportunities for collective action outside of the organization.

The performance assessment construct from OIT put a spotlight on current outcomes to compare against each program's goals. We used diverse metrics as indicators of outcomes outside the control of the zoo to measure their progress. With monitoring the 1833 endangered beetles and 1400 plants reintroduced to the wild, 2979 gardens created, 350 children engaged, 300 native plant species, and 12,522 individual plants sold, and over 200 new species spotted in restored habitats, the programs are making progress toward their primary goals set as intentions for desired outcomes (Table 1). The program teams showed they are actively documenting the animals and plants reintroduced and observed, as well as the communities mobilized to participate. We are making progress for insects and plant conservation efforts in our communities, as represented by the widespread projects in our region (Figure 1). The five case studies highlight CZBG's leadership in local flora and fauna conservation with a range of techniques and outcome metrics monitoring the species and landscapes (Table 1). These programs are ongoing, as we have yet to completely achieve the program goals; therefore, CZBG plans to use the opportunities for improving the projects highlighted by the OIT framework. For example, Bowyer Farm, Plant for Pollinators, and the Rockdale Urban Learning Garden do not have elevating the zoo as a conservation leader as a priority or secondary goal. While these projects do important work, they are not well known beyond the immediate communities. The OIT evaluation highlights which projects need better storytelling and community activation, as well as which can diversify their impact to achieve CZBG's organizational coexistence goals of helping both wildlife and people to thrive.

Future adaptive management of the programs will reflect on ways to better mobilize new participants—both internal zoo staff and external zoo visitors—as well as facilitating justice (e.g., alleviating financial barriers to individuals participating in Plant for Pollinators). More success could be found in local projects by increasing the coexistence goal of justice, such as by engaging local underrepresented experts through cooperative partnerships. Rockdale ULG is one example of a project using the coexistence goal of justice effectively. Goals stated by the Rockdale Academy Principal Jaren Finney include using the ULG to teach students "about agriculture, horticulture, and even how to be entrepreneurs" [76]. This active partnership with local diverse education experts that know and represent their community guides the way CZBG empowers their needs with equitable resources and inclusive communications. Other zoos and botanical gardens can reflect on how to integrate such methods from this justice-focused case study.

5. Conclusions

Native plant and insect projects at Cincinnati Zoo & Botanical Garden are fueled by collaborative partnerships as they work toward multifaceted coexistence goals. Through this evaluation, we show current commitments to native wildlife conservation and present opportunities for adaptive management to further enhance the activation of a clear and effective organizational conservation identity. The zoos' conservation programs grounded in local landscapes have opportunities to mobilize internal staff and external partners throughout their communities. Furthermore, programs with goals and activities supporting wildlife, community involvement, and environmental justice enable stronger conservation identities. The results of this study highlight how other zoos and botanical gardens can lead the conservation of local flora and fauna in their communities with the reflective evaluation process. Moreover, the internal collaborative evaluation methods using the OIT framework highlight components critical to strategy implementation and best practices relevant to other zoos, aquariums, and botanical gardens.

Author Contributions: Conceptualization, L.M., B.C., T.T.; methodology, L.M., B.C., T.T.; formal analysis, L.M., B.C.; writing—original draft preparation, L.M., B.C., T.T.; writing—review and editing, V.P., M.P. (Megan Philpott), M.O., M.P. (Mandy Pritchard), J.G., B.R., J.H., D.J.; project administration, V.P., M.P. (Megan Philpott), M.O., M.P. (Mandy Pritchard), J.G., D.J.; funding acquisition, L.M., D.J. All authors have read and agreed to the published version of the manuscript.

Funding: Analysis and paper writing has been supported by the Cincinnati Zoo & Botanical Garden.

Institutional Review Board Statement: Ethical review and approval were waived for this study due to the internal evaluation structure of this study that involved project partners in the collaborative evaluation through group discussion. As such, most participants in the study are co-authors on this paper or acknowledged below. However, the research team followed best practices in accordance with the Declaration of Helsinki, by obtaining informed consent from all subjects and reminding each participant at each meeting that the discussions were voluntary and their identity would be anonymous in the data collection and results reporting.

Informed Consent Statement: Informed consent was obtained from all subjects involved in the study.

Data Availability Statement: The data presented in this study are available on request from the corresponding author, subject to permission from study partners.

Acknowledgments: The authors are grateful for all the project partners and Cincinnati Zoo & Botanical Garden staff, visitors, and donors that have supported these five case study programs. We would like to thank Steve Foltz, Scott Beuerlein, Jerome Stenger, Brian Jorg, Chase Bodkin, Carlos VanLeeuwen, Christina Anderson, and Winton Ray for their insight into and leadership of the CZBG programs highlighted in this paper.

Conflicts of Interest: The authors declare no conflict of interest.

References

1. Cardoso, P.; Barton, P.S.; Birkhofer, K.; Chichorro, F.; Deacon, C.; Fartmann, T.; Fukushima, C.S.; Gaigher, R.; Habel, J.C.; Hallmann, C.A.; et al. Scientists' warning to humanity on insect extinctions. *Biol. Conserv.* **2020**, *242*, 108426. [CrossRef]
2. Goulson, D. The insect apocalypse, and why it matters. *Curr. Biol.* **2019**, *29*, R967–R971. [CrossRef] [PubMed]
3. Sánchez-Bayo, F.; Wyckhuys, K. Worldwide decline of the entomofauna: A review of its drivers. *Biol. Conserv.* **2019**, *232*, 8–27. [CrossRef]
4. Van Klink, R.; Bowler, D.E.; Gongalsky, K.B.; Swengel, A.B.; Gentile, A.; Chase, J.M. Meta-analysis reveals declines in terrestrial but increases in freshwater insect abundances. *Science* **2020**, *368*, 417–420. [CrossRef]
5. Piano, E.; Souffreau, C.; Merckx, T.; Baardsen, L.F.; Backeljau, T.; Bonte, D.; Brans, K.I.; Cours, M.; Dahirel, M.; Debortoli, N.; et al. Urbanization drives cross-taxon declines in abundance and diversity at multiple spatial scales. *Glob. Change Biol.* **2020**, *26*, 1196–1211. [CrossRef]
6. Maynard, L.; Jacobson, S.K.; Monroe, M.; Savage, A. Mission impossible or Mission Accomplished? Do zoos' organizational missions influence their conservation practices? *Zoo Biol.* **2020**, *39*, 304–314. [CrossRef]
7. Belsky, J.; Joshi, N.K. Assessing Role of Major Drivers in Recent Decline of Monarch Butterfly Population in North America. *Front. Environ. Sci.* **2018**, *6*, 86. [CrossRef]
8. Crone, E.E.; Pelton, E.M.; Brown, L.M.; Thomas, C.C.; Schultz, C.B. Why are monarch butterflies declining in the West? Understanding the importance of multiple correlated drivers. *Ecol. Appl.* **2019**, *29*, e01975. [CrossRef]
9. Pelton, E.M.; Schultz, C.B.; Jepsen, S.J.; Black, S.H.; Crone, E.E. Western Monarch Population Plummets: Status, Probable Causes, and Recommended Conservation Actions. *Front. Ecol. Evol.* **2019**, *7*, 258. [CrossRef]
10. Simaika, J.P.; Samways, M.J. Insect conservation psychology. *J. Insect Conserv.* **2018**, *22*, 635–642. [CrossRef]
11. Elmqvist, T.; Folke, C.; Nyström, M.; Peterson, G.; Bengtsson, J.; Walker, B.; Norberg, J. Response Diversity, Ecosystem Change, and Resilience. *Front. Ecol. Environ.* **2003**, *1*, 488–494. [CrossRef]
12. Reilly, J.R.; Artz, D.R.; Biddinger, D.; Bobiwash, K.; Boyle, N.K.; Brittain, C.; Brokaw, J.; Campbell, J.W.; Daniels, J.; Elle, E.; et al. Crop production in the USA is frequently limited by a lack of pollinators. *Proc. R. Soc. B Boil. Sci.* **2020**, *287*, 20200922. [CrossRef]
13. Wagner, D.L.; Grames, E.M.; Forister, M.L.; Berenbaum, M.R.; Stopak, D. Insect decline in the Anthropocene: Death by a thousand cuts. *Proc. Natl. Acad. Sci. USA* **2021**, *118*, e2023989118. [CrossRef]
14. Antonelli, A.; Smith, R.J.; Fry, C.; Simmonds, M.S.; Kersey, P.J.; Pritchard, H.W.; Abbo, M.S.; Acedo, C.; Adams, J.; Ainsworth, A.M.; et al. *State of the World's Plants and Fungi 2020*; Royal Botanic Gardens, Kew: Richmond, UK, 2020. [CrossRef]
15. Balding, M.; Williams, K.J. Plant blindness and the implications for plant conservation. *Conserv. Biol.* **2016**, *30*, 1192–1199. [CrossRef]
16. Negrón-Ortiz, V. Pattern of expenditures for plant conservation under the Endangered Species Act. *Biol. Conserv.* **2014**, *171*, 36–43. [CrossRef]
17. Oldfield, S.F. Botanic gardens and the conservation of tree species. *Trends Plant Sci.* **2009**, *14*, 581–583. [CrossRef]
18. Rupprecht, C.D.; Byrne, J.A.; Garden, J.G.; Hero, J.-M. Informal urban green space: A trilingual systematic review of its role for biodiversity and trends in the literature. *Urban For. Urban Green.* **2015**, *14*, 883–908. [CrossRef]
19. Milesi, C.; Running, S.W.; Elvidge, C.D.; Dietz, J.B.; Tuttle, B.T.; Nemani, R.R. Mapping and Modeling the Biogeochemical Cycling of Turf Grasses in the United States. *Environ. Manag.* **2005**, *36*, 426–438. [CrossRef]
20. Allen, W.; Ballmori, D.; Haeg, F. *Edible Estates: Attack on the Front Lawn*; Metropolis Books: New York, NY, USA, 2010.
21. U.S. Census Bureau. *Statistical Abstract of the United States: 2010*, 129th ed.; U.S. Census Bureau: Washington, DC, USA, 2021. Available online: http://www.census.gov/compendia/statab/ (accessed on 15 August 2022).
22. Forister, M.L.; Pelton, E.M.; Black, S.H. Declines in insect abundance and diversity: We know enough to act now. *Conserv. Sci. Pract.* **2019**, *1*, e80. [CrossRef]
23. Campbell, J.W.; Kimmel, C.B.; Grodsky, S.M.; Smithers, C.; Daniels, J.C.; Ellis, J.D. Wildflower plantings harbor increased arthropod richness and abundance within agricultural areas in Florida (USA). *Ecosphere* **2019**, *10*, e02890. [CrossRef]
24. Shuey, J.A. Habitat Re-Creation (Ecological Restoration) as a Strategy for Conserving Insect Communities in Highly Fragmented Landscapes. *Insects* **2013**, *4*, 761–780. [CrossRef] [PubMed]
25. Adams, B.J.; Li, E.; Bahlai, C.A.; Meineke, E.K.; McGlynn, T.P.; Brown, B.V. Local- and landscape-scale variables shape insect diversity in an urban biodiversity hot spot. *Ecol. Appl.* **2020**, *30*, e02089. [CrossRef] [PubMed]

26. Baldock, K.C.R.; Goddard, M.; Hicks, D.M.; Kunin, W.E.; Mitschunas, N.; Osgathorpe, L.M.; Potts, S.G.; Robertson, K.M.; Scott, A.V.; Stone, G.; et al. Where is the UK's pollinator biodiversity? The importance of urban areas for flower-visiting insects. *Proc. R. Soc. B Boil. Sci.* **2015**, *282*, 20142849. [CrossRef] [PubMed]
27. Baldock, K.C.R.; Goddard, M.A.; Hicks, D.M.; Kunin, W.E.; Mitschunas, N.; Morse, H.; Osgathorpe, L.M.; Potts, S.G.; Robertson, K.M.; Scott, A.V.; et al. A systems approach reveals urban pollinator hotspots and conservation opportunities. *Nat. Ecol. Evol.* **2019**, *3*, 363–373. [CrossRef]
28. Luong, J.C.; Turner, P.L.; Phillipson, C.N.; Seltmann, K.C. Local grassland restoration affects insect communities. *Ecol. Èntomol.* **2019**, *44*, 471–479. [CrossRef]
29. Theodorou, P.; Albig, K.; Radzevičiūtė, R.; Settele, J.; Schweiger, O.; Murray, T.E.; Paxton, R.J. The structure of flower visitor networks in relation to pollination across an agricultural to urban gradient. *Funct. Ecol.* **2016**, *31*, 838–847. [CrossRef]
30. Gill, K.A.; Cox, R.; O'Neal, M.E. Quality Over Quantity: Buffer Strips Can be Improved With Select Native Plant Species. *Environ. Èntomol.* **2014**, *43*, 298–311. [CrossRef]
31. Mathiasson, M.E.; Rehan, S.M. Wild bee declines linked to plant-pollinator network changes and plant species introductions. *Insect Conserv. Divers.* **2020**, *13*, 595–605. [CrossRef]
32. Potts, S.G.; Biesmeijer, J.C.; Kremen, C.; Neumann, P.; Schweiger, O.; Kunin, W.E. Global pollinator declines: Trends, impacts and drivers. *Trends Ecol. Evol.* **2010**, *25*, 345–353. [CrossRef]
33. Schultz, C.; Russell, C.; Wynn, L. Restoration, Reintroduction, and captive Propagation for at-risk Butterflies: A review of British and American Conservation Efforts. *Isr. J. Ecol. Evol.* **2008**, *54*, 41–61. [CrossRef]
34. Braatz, E.Y.; Gezon, Z.J.; Rossetti, K.; Maynard, L.T.; Bremer, J.S.; Hill, G.M.; Streifel, M.A.; Daniels, J.C. Bloom evenness modulates the influence of bloom abundance on insect community structure in suburban gardens. *PeerJ* **2021**, *9*, e11132. [CrossRef]
35. Lewis, A.D.; Bouman, M.J.; Winter, A.M.; Hasle, E.A.; Stotz, D.F.; Johnston, M.K.; Klinger, K.R.; Rosenthal, A.; Czarnecki, C.A. Does Nature Need Cities? Pollinators Reveal a Role for Cities in Wildlife Conservation. *Front. Ecol. Evol.* **2019**, *7*, 220. [CrossRef]
36. Maynard, L.; Howorth, P.; Daniels, J.; Bunney, K.; Snyder, R.; Jenike, D.; Barnhart, T.; Spevak, E.; Fitzgerald, P.; Gezon, Z. Conservation psychology strategies for collaborative planning and impact evaluation. *Zoo Biol.* **2022**, *41*, 425–438. [CrossRef]
37. Bailey, K.; Morales, N.; Newberry, M. Inclusive conservation requires amplifying experiences of diverse scientists. *Nat. Ecol. Evol.* **2020**, *4*, 1294–1295. [CrossRef]
38. Alvarez, I.; Lovera, S. New Times for Women and Gender Issues in Biodiversity Conservation and Climate Justice. *Development* **2016**, *59*, 263–265. [CrossRef]
39. Austin, J.E.; Seitanidi, M.M. Collaborative Value Creation: A Review of Partnering Between Nonprofits and Businesses. Part 2: Partnership Processes and Outcomes. *Nonprofit Volunt. Sect. Q.* **2012**, *41*, 929–968. [CrossRef]
40. Maynard, L.; Mccarty, C.; Jacobson, S.K.; Monroe, M.C. Conservation networks: Are zoos and aquariums collaborating or competing through partnerships? *Environ. Conserv.* **2020**, *47*, 166–173. [CrossRef]
41. Che-Castaldo, J.P.; Grow, S.A.; Faust, L.J. Evaluating the Contribution of North American Zoos and Aquariums to Endangered Species Recovery. *Sci. Rep.* **2018**, *8*, 9789. [CrossRef]
42. Miller, B.; Conway, W.; Reading, R.P.; Wemmer, C.; Wildt, D.; Kleiman, D.; Monfort, S.; Rabinowitz, A.; Armstrong, B.; Hutchins, M. Evaluating the Conservation Mission of Zoos, Aquariums, Botanical Gardens, and Natural History Museums. *Conserv. Biol.* **2004**, *18*, 86–93. [CrossRef]
43. Black, S.A. A Leadership Competence Framework to Support the Development of Conservation Professionals. *Open J. Leadersh.* **2021**, *10*, 300–337. [CrossRef]
44. Bruyere, B.L. Giving Direction and Clarity to Conservation Leadership. *Conserv. Lett.* **2015**, *8*, 378–382. [CrossRef]
45. Armsworth, P.; Larson, E.; Boyer, A. Adaptability: As important in conservation organizations as it is in species. In *Effective Conservation Science: Data Not Dogma*; Oxford University Press: Oxford, UK, 2017; pp. 58–63. [CrossRef]
46. Salerno, J.; Romulo, C.; A Galvin, K.; Brooks, J.; Mupeta-Muyamwa, P.; Glew, L. Adaptation and evolution of institutions and governance in community-based conservation. *Conserv. Sci. Pract.* **2021**, *3*, e355. [CrossRef]
47. Rastogi, A.; Thapliyal, S.; Hickey, G.M. Community Action and Tiger Conservation: Assessing the Role of Social Capital. *Soc. Nat. Resour.* **2014**, *27*, 1271–1287. [CrossRef]
48. Berkes, F. Rethinking Community-Based Conservation. *Conserv. Biol.* **2004**, *18*, 621–630. [CrossRef]
49. Wilkins, K.; Pejchar, L.; Carroll, S.L.; Jones, M.S.; Walker, S.E.; Shinbrot, X.A.; Huayhuaca, C.; Fernández-Giménez, M.E.; Reid, R.S. Collaborative conservation in the United States: A review of motivations, goals, and outcomes. *Biol. Conserv.* **2021**, *259*, 109165. [CrossRef]
50. Parris, D.L.; Peachey, J.W. Encouraging servant leadership: A qualitative study of how a cause-related sporting event inspires participants to serve. *Leadership* **2013**, *9*, 486–512. [CrossRef]
51. Silvertown, J. A new dawn for citizen science. *Trends Ecol. Evol.* **2009**, *24*, 467–471. [CrossRef]
52. Lucas, J.; Gora, E.; Alonso, A. A view of the global conservation job market and how to succeed in it. *Conserv. Biol.* **2017**, *31*, 1223–1231. [CrossRef]
53. Ardoin, N.M.; Bowers, A.W.; Gaillard, E. Environmental education outcomes for conservation: A systematic review. *Biol. Conserv.* **2020**, *241*, 108224. [CrossRef]
54. Kullenberg, C.; Kasperowski, D. What is citizen science? A scientometric meta-analysis. *PLoS ONE* **2016**, *11*, e0147152. [CrossRef]

55. Vann-Sander, S.; Clifton, J.; Harvey, E. Can citizen science work? Perceptions of the role and utility of citizen science in a marine policy and management context. *Mar. Policy* **2016**, *72*, 82–93. [CrossRef]
56. Theobald, E.; Ettinger, A.; Burgess, H.; DeBey, L.; Schmidt, N.; Froehlich, H.; Wagner, C.; HilleRisLambers, J.; Tewksbury, J.; Harsch, M.; et al. Global change and local solutions: Tapping the unrealized potential of citizen science for biodiversity research. *Biol. Conserv.* **2015**, *181*, 236–244. [CrossRef]
57. Toomey, A.H.; Knight, A.T.; Barlow, J. Navigating the Space between Research and Implementation in Conservation. *Conserv. Lett.* **2016**, *10*, 619–625. [CrossRef]
58. Cook, C.N.; Mascia, M.B.; Schwartz, M.W.; Possingham, H.P.; Fuller, R.A. Achieving Conservation Science that Bridges the Knowledge–Action Boundary. *Conserv. Biol.* **2013**, *27*, 669–678. [CrossRef]
59. Albert, S.; Whetten, D.A. Organizational identity. *Res. Organ. Behav.* **1985**, *7*, 263–295.
60. King, B.G.; Felin, T.; Whetten, D.A. Perspective—Finding the Organization in Organizational Theory: A Meta-Theory of the Organization as a Social Actor. *Organ. Sci.* **2010**, *21*, 290–305. [CrossRef]
61. Whetten, D.A.; Mackey, A. A Social Actor Conception of Organizational Identity and Its Implications for the Study of Organizational Reputation. *Bus. Soc.* **2002**, *41*, 393–414. [CrossRef]
62. He, H.; Balmer, J.M. Identity studies: Multiple perspectives and implications for corporate-level marketing. *Eur. J. Mark.* **2007**, *41*, 765–785. [CrossRef]
63. Melewar, T.C.; Karaosmanoglu, E.; Paterson, D. Corporate identity: Concept, components and contribution. *J. Gen. Manag.* **2005**, *31*, 59–81. [CrossRef]
64. Maynard, L.; Adams, A.E.; Jacobson, S.K.; Monroe, M.C. Evaluating Organizational Identity of Zoos to Enhance Conservation. *Curator Mus. J.* **2021**, *64*, 549–565. [CrossRef]
65. Wahlén, C. Constructing Conservation Impact: Understanding Monitoring and Evaluation in Conservation NGOs. *Conserv. Soc.* **2014**, *12*, 77. [CrossRef]
66. Fraser, J.; Wharton, D. The Future of Zoos: A New Model for Cultural Institutions. *Curator Mus. J.* **2007**, *50*, 41–54. [CrossRef]
67. Solis-Sosa, R.; Semeniuk, C.A.; Fernandez-Lozada, S.; Dabrowska, K.; Cox, S.; Haider, W. Monarch butterfly conservation through the social lens: Eliciting public preferences for management strategies across transboundary nations. *Front. Ecol. Evol.* **2019**, *7*, 316. [CrossRef]
68. Census Reporter. Cincinnati, OH-KY-IN Metro Area. 2021. Available online: https://censusreporter.org/profiles/31000US17140-cincinnati-oh-ky-in-metro-area/ (accessed on 15 August 2022).
69. Ashforth, B.E.; Harrison, S.H.; Corley, K.G. Identification in Organizations: An Examination of Four Fundamental Questions. *J. Manag.* **2008**, *34*, 325–374. [CrossRef]
70. Margolis, S.L.; Hansen, C.D. A Model for Organizational Identity: Exploring the Path to Sustainability during Change. *Hum. Resour. Dev. Rev.* **2002**, *1*, 277–303. [CrossRef]
71. Tlili, A. The organisational identity of science centres. *Cult. Organ.* **2008**, *14*, 309–323. [CrossRef]
72. Saldana, J. *The Coding Manual for Qualitative Researchers*; SAGE Publishers: Los Angeles, CA, USA, 2016.
73. Kjærgaard, A.L. Organizational Identity and Strategy: An Empirical Study of Organizational Identity's Influence on the Strategy-Making Process. *Int. Stud. Manag. Organ.* **2009**, *39*, 50–69. [CrossRef]
74. Pence, V.C.; Plair, B.L.; Charls, S.M.; Clark, J.R.; Taylor, D.D. Micropropagation, Cryopreservation, and Outplanting of the Cumberland Sandwort Minuartia cumberlandensis. *J. Ky. Acad. Sci.* **2011**, *72*, 91–99. [CrossRef]
75. Mckenna-Foster, A.; Perrotti, L. Re-introduction of the American burying beetle to Nantucket Island, Massachusetts, USA. In *Global Re-Introduction Perspectives: 2011: More Case Studies from Around the Globe*; Soorae, P.S., Ed.; IUCN/SSC Re-Introduction Specialist Group & Environmental Agency: Brussels, Belgium, 2011; pp. 1–4.
76. Sharber, C. New Urban Learning Garden at Rockdale Academy Could be 'Heart of the Community' in Avondale. [Radio Broadcast]. WVXU. (30 September 2021). Available online: https://www.wvxu.org/community/2021-09-30/rockdale-academy-urban-learning-garden (accessed on 12 August 2022).

MDPI
St. Alban-Anlage 66
4052 Basel
Switzerland
Tel. +41 61 683 77 34
Fax +41 61 302 89 18
www.mdpi.com

Journal of Zoological and Botanical Gardens Editorial Office
E-mail: jzbg@mdpi.com
www.mdpi.com/journal/jzbg

www.ingramcontent.com/pod-product-compliance
Lightning Source LLC
Chambersburg PA
CBHW040717270926
51850CB00023B/2675

9783036578651